COMPUTER NETWORKS

SYNCHRONIZATION IN COMPLEX NETWORKS

COMPUTER NETWORKS

Additional books in this series can be found on Nova's website under the Series tab.

Additional E-books in this series can be found on Nova's website under the E-books tab.

COMPUTER SCIENCE, TECHNOLOGY AND APPLICATIONS

Additional books in this series can be found on Nova's website under the Series tab.

Additional E-books in this series can be found on Nova's website under the E-books tab.

COMPUTER NETWORKS

SYNCHRONIZATION IN COMPLEX NETWORKS

XIN BIAO LU
AND
BU ZHI QIN

Nova Science Publishers, Inc.
New York

Copyright © 2011 by Nova Science Publishers, Inc.

All rights reserved. No part of this book may be reproduced, stored in a retrieval system or transmitted in any form or by any means: electronic, electrostatic, magnetic, tape, mechanical photocopying, recording or otherwise without the written permission of the Publisher.

For permission to use material from this book please contact us:
Telephone 631-231-7269; Fax 631-231-8175
Web Site: http://www.novapublishers.com

NOTICE TO THE READER

The Publisher has taken reasonable care in the preparation of this book, but makes no expressed or implied warranty of any kind and assumes no responsibility for any errors or omissions. No liability is assumed for incidental or consequential damages in connection with or arising out of information contained in this book. The Publisher shall not be liable for any special, consequential, or exemplary damages resulting, in whole or in part, from the readers' use of, or reliance upon, this material. Any parts of this book based on government reports are so indicated and copyright is claimed for those parts to the extent applicable to compilations of such works.

Independent verification should be sought for any data, advice or recommendations contained in this book. In addition, no responsibility is assumed by the publisher for any injury and/or damage to persons or property arising from any methods, products, instructions, ideas or otherwise contained in this publication.

This publication is designed to provide accurate and authoritative information with regard to the subject matter covered herein. It is sold with the clear understanding that the Publisher is not engaged in rendering legal or any other professional services. If legal or any other expert assistance is required, the services of a competent person should be sought. FROM A DECLARATION OF PARTICIPANTS JOINTLY ADOPTED BY A COMMITTEE OF THE AMERICAN BAR ASSOCIATION AND A COMMITTEE OF PUBLISHERS.

Additional color graphics may be available in the e-book version of this book.

LIBRARY OF CONGRESS CATALOGING-IN-PUBLICATION DATA
Lu, Xin Biao.
Synchronization in complex networks / authors, Xin Biao Lu, Bu Zhi Qin.
p. cm.
Includes index.
ISBN 978-1-61761-873-4 (softcover)
1. Synchronous data transmission systems. 2. Computer network architectures. I. Qin, Bu Zhi. II. Title.
TK5105.415.L825 2011
004.6'5--dc22
2010034047

Published by Nova Science Publishers, Inc. † New York

Contents

Preface vii

Chapter 1	Summarization of Synchronization in Complex Network	1
Chapter 2	Adaptive Synchronization of Complex Networks	23
Chapter 3	Cluster Synchronization in Complex Networks	41
Chapter 4	Control of Complex Dynamical Networks	67
Chapter 5	Synchronization of Time Varying Complex Networks	93

Acknowledgments 121

References 123

Index 135

PREFACE

As the improvement of computer's calculation capacity and the rapid development of information technology, people began to investigate various dynamical behaviors on the complex network with lots of nodes and complex network topology. As one of these dynamical behaviors, synchronization of complex networks has been a hot research topic.

In this book, the latest developments of synchronization of complex networks such as the adaptive synchronization, cluster synchronization of complex networks with fixed topologies; and synchronization of complex network with switched topologies.

We would like to gratefully acknowledge the following publishers for giving us permissions to use figures that appeared in past publications: The Institute of Electrical and Electronic Engineers (IEEE), Elsevier, American Physical Society, American Institute of Physics, Communications in Theoretical Physics and SIAM. Wherever a figure from a past publication, we give the full reference to the original paper, and indicate in the caption of the corresponding figure that the copyright belongs to the appreciate publishers (for example, "© IEEE").

We would like to acknowledge the financial support from the Fundamental Research Funds for the Central Universities under Grant No. 2009B20114 and the Natural Science Foundation for Young Scholars of Nanjing College of Chemical Technology.

Xin Biao Lu
Hohai University
Nanjing, China
October, 2010.

Chapter 1

SUMMARIZATION OF SYNCHRONIZATION IN COMPLEX NETWORK

ABSTRACT

The summarization of synchronization in a complex network is introduced in this chapter. At first, the basic concepts of a complex network, including the description of the network, the degree of the node, clustering coefficient, the average path length, and betweenness centrality are introduced. Then, the definition of synchronization in a complex network and its stability analysis methods are presented. When the initial states of nodes are near enough to synchronization manifold, the master stability function method is applied to analyze its local stability. However, when the initial states of nodes are randomly distributed, the Lyapunov function method is used to analyze the global stability of synchronization manifold. Furthermore, the connection graph stability method is used to investigate the global stability of synchronization in a complex network with time-varying network topology.

1.1. INTRODUCTION

Synchronization implies that the states of two or more interacting agents with different initial conditions gradually approach each other as the time increases; finally, all the agents reach the same state. Synchronization phenomenon is ubiquitous in nature. In 1665, the famous physicist C.

Huygens, the inventor of the pendulum clock, discovered two pendulum clocks, hung on a sickroom wall, swing with exactly the same frequency irrespective of different initial phases. Furthermore, he pointed out that the reason of these two pendulum clocks reaching synchronization is that the interaction came from the common beam supporting the two clocks [1]. Synchronization is often seen in the daily life, For example: a group of frogs in the pond often ring together after rain on the summer night[2]; the fireflies on the same tree synchronously light and synchronously do not light; after a wonderful perform, at first the applause of the audience is disorderly and unsystematic, while after several seconds, the rhythm of applause of the audience becomes same [3, 4]. Recently, it was found that synchronization in heart cell network and brain neural network [5, 6]; even the nano-oscillators also happen to synchronize with each other by mutual coupling strength between them[7, 8].

On the contrary to the above helpful synchronization behavior, synchronization behavior may be baneful. For example, on the ceremony of the London millenary bridge, when thousands of people began to cross the bridge, the maximum warp caused by the resonance was near to 20 centimeters, the people on the bridge became rattled. Therefore, the bridge had to be closed temporarily.

How to describe the synchronization behavior of coupled dynamical oscillators has been interested by researchers from multiple domains such as biology, technology and sociology for a long time [9, 10]. Winfree inaugurated the research of synchronization of interaction oscillators [11, 12]. He supposed that each oscillator only has strong connection with its surrounding limited oscillators, the variation of the oscillators' amplitude may be neglected, therefore the synchronization problem of interacting oscillators reduces to the problem to solve the variation of the oscillator' phase. Under the assumption, a group of non-identical oscillators with intrinsic frequencies that were distributed about some mean value, according to some prescribed probability distribution, may exhibit a remarkable synchronization behavior. When the variance of the frequency distribution is larger than a threshold, the oscillators run incoherently. However, when the variance is lower than the threshold, the oscillators become to synchronize spontaneously [12]. It is worth to mention that the origin Winfree model was only simulation results. Recently, Ariaratnam gave the mathematical analysis for the origin Winfree model [13].

The Winfree model is well to describe the emergence of spontaneous order in the oscillators, but it was based on the assumption that each oscillator is connected. This all-to-all connectivity between oscillators is difficult to be applied when the number of oscillators is large enough, considering the optimization of cost. Therefore, it is necessary to modify the Winfree model. In 1998, Watts and Strogatz gave a small-world network model [14]. Starting with a regular lattice, it was found that by adding few random links, the distance between nodes is reduced drastically, which is known as small-world (SW). Since the new model is successful in describing the transition from regular lattices and random graphs, the SW network model proposed by Watts and Strogatz is called the WS model. The WS model has a high degree of clustering as in the regular networks and a small average path length among nodes as in the random networks. Both random graph and WS small-world network have exponential degree distribution, which means that these networks are homogeneous in the sense that all nodes have roughly the same number of connections. In 1999, Barabasi and Albert found that most complex networks such as the Internet, the World-Wide Web and scientific collaboration networks have a heavy tailed distribution of connectivity with no characteristic scale, which is the season to be called the scale free (SF) network [15]. The SF network has a power-law degree distribution $p(k) \sim k^{-\gamma}$ where $p(k)$ is the probability that a node in the network is connected to k other nodes and γ is a positive number. These scale-free networks are heterogeneous, in which most nodes have very few connections but a small number of particular nodes have many connections. The known SF network model proposed by Barabasi and Albert is represented by BA network model.

The observation of these two features prompts more and more scholars to investigate the connection of network topology and dynamical behaviors of complex network. As one of important behaviors of complex network, synchronization of lots of oscillators has been a hot research topic, especially for the past several years. The main goal of this book is to survey the affect of the interaction of network topology and the oscillator's dynamics on the synchronization of complex network with many oscillators.

1.2. BASIC CONCEPT OF NETWORK

The detailed definition and explain of the fundamental parameters may refer to the known books and reviews [16-22]. In order to make the reader understand the book easily, the main notations used in the throughout book are introduced in brief.

1.2.1. The Graph Description of Network

A complex network may be described by a graph G made up of a set of N nodes (or vertices) connected by a set of M links (or edges). The degree of node i is the number of links between node i and its neighbors. If any pair (i,j) and (j,i) correspond to one same edge, then the network is undirected; otherwise, the network is directed. If each edge is assigned a corresponding strength, then the network is weighted, otherwise, the network is un-weighted. The graph of the un-weighted network is represented by the adjacency matrix A, with entries $a_{ij}=1$ if a directed link from node j to node i exists, and 0 otherwise. The graph of the weighted network is represented by a matrix W, with entries w_{ij}, representing the strength (or weight) of the link from node j to node i.

1.2.2. The Average Path Length

The distance between node j to node i is defined as the number of the shortest path between these two nodes. The maximum of the distance between any two nodes in the network is the network's diameter D, which means that

$$D = \max_{i,j} d_{ij} \qquad (1\text{-}1)$$

The average path length is the average algorithmic sum of the distance between random two nodes, that is

$$L = \frac{1}{\frac{1}{2}N(N+1)} \sum_{i \geq j} d_{ij} \qquad (1\text{-}2)$$

where N is the number of the whole nodes in the network.

1.2.3. Clustering Coefficient

The clustering property describes the phenomenon that one node's neighbors may be neighbors each other. The clustering coefficient of one node is usually defined as follows:

$$C_i = 2E_i / (k_i(k_i - 1)) \qquad (1\text{-}3)$$

where E_i is the number of connections between node i and its nearest neighbors, and k_i is its degree. A large clustering coefficient means that there exist many triangles between node i and its neighbors, while a low clustering coefficient implies that there exist few triangles between them.

The clustering coefficient of the network is the average algorithmic sum of every node's clustering coefficient.

1.2.4. Betweenness

The information exchange between two non-adjacent nodes, node j and node k, depends on the existing paths connecting them. Node betweenness is a given node may be represented by the number of shortest paths between them in the network throughout it. Furthermore, the betweenness b_i of node i, sometimes referred to as load, is defined as

$$b_i = \sum_{j,k \in N, j \neq k} \frac{n_{jk}(i)}{n_{jk}} \qquad (1\text{-}4)$$

where n_{jk} is the number of shortest paths connecting node j and node k, while $n_{jk}(i)$ is the number of shortest paths connecting node j and node k and passing through node i [22].

The above node betweenness may be extended to the edge betweenness, which is defined as the number of shortest paths between pairs of nodes through the edge.

1.2.5. Assortative Coefficient

The assortative coefficient is used to describe the connection orientation between one node and the other nodes. Its definition is as follows [23]

$$r = \frac{\sum_i j_i k_i - M^{-1} \sum_i j_i \sum_i k_i}{\sqrt{\left[\sum_i j_i^2 - M^{-1}(\sum_i j_i)^2\right]\left[\sum_i k_i^2 - M^{-1}(\sum_i k_i)^2\right]}} \quad (1\text{-}5)$$

where j_i and k_i are the in-degree and out-degree of the nodes at the ends of the i th edge, respectively; M is the total number of edges in the network. For the undirected network, one undirected edge may be replaced by two directed edges.

When the assortative coefficient is bigger than 0, the network is assortative, which implies that the node with higher/lower degree easy to connect with node with higher/lower degree. Otherwise, when the assortative coefficient is less than 0, the network is dis-assortative, which implies that the node with higher/lower degree easy to connect with node with lower/higher degree.

1.3. COMPLETE SYNCHRONIZATION IN COMPLEX NETWORK

Recently, complete synchronization of complex network with interacting chaotic system has been a hot research topic [20, 24]. As the simplest form of synchronization, complete synchronization (or identical synchronization) is perfect to investigate the trajectories of identical chaotic systems evolving in the course of the time. Since the complete synchronization of complex networks with scale-free or small-world topologies may be analyzed, it is interested by most researchers. In the latter book, the synchronization of complex network with general coupling configuration is given.

1.3.1. Master Stability Function

Barahona and Pecora initiated the research of the complete synchronization of complex networks with small-world topology by proposing a master stability function (MSF) [25]. After a short time, the MSF method is extended to the case of complex networks with more general topologies [26-30].

Consider a dynamical network of N linearly coupled identical oscillators, with each oscillator being an n-dimensional dynamical system. Let each oscillator of the network be assigned a dynamical variable $x_i (i=1,2,\cdots,N)$. The evolution of the dynamical variables is written in the form [31-35]:

$$\dot{x}_i(t) = f(x_i(t)) + c\sum_{j=1}^{N} a_{ij} H(x_j(t)), \qquad (1\text{-}6)$$

where the function f describes the dynamics of each individual oscillator, H is the inner coupling matrix, c is the coupling strength between nodes, the coupling matrix $A = (a_{ij}) \in R^{N \times N}$ represents the network topology: If there is a connection between node i and node j, then $a_{ij} > 0$; otherwise, $a_{ij} = 0$; furthermore, the dissipation condition $\sum_j a_{ij} = 0$ satisfies. This implies that the coupling matrix A has zero row-sum. A complete synchronization state exists, that is

$$x_1(t) \to x_2(t) \to \cdots \to x_N(t) \to s(t) \qquad (1\text{-}7)$$

If the time $t \to \infty$, the equation (1-7) satisfies, then the network is said to achieve complete synchronization. The synchronization state $s(t)$ must be a solution of an isolated node, that is $\dot{s}(t) = f(s(t))$, where $s(t)$ may be equilibrium point, periodic trajectory and even chaotic trajectory. The substance (1-7) is called synchronization manifold, where all the nodes oscillators synchronously evolve on the same solution of the isolated node.

Obviously, when all the nodes are initially assigned at the synchronization manifold, the synchronization may be remain (Here I think the word should be "remained", since it is passiveness). However, it is meaningful to investigate the effect of small perturbations on the synchronization manifold. By assigning $x_i(t) = s(t) + \xi(t)$, where $\xi(t)$ is a small perturbation, then the equation (1-6) becomes

$$\dot{\xi}(t) = f(x_i(t)) - f(s(t)) + c\sum_{j=1}^{N} a_{ij} H(x_j(t)) - c\sum_{j=1}^{N} a_{ij} H(s(t)), \qquad (1\text{-}8)$$

Since $\dot{s}(t) = f(s(t))$, then $-c\sum_{j=1}^{N} a_{ij} H(s(t))$ is zero. Expand the functions f and H near the synchronization state to first order in a Taylor series, that is

$$f(x_i(t)) = f(s(t)) + Df(s(t))\xi$$
$$H(x_i(t)) = H(s(t)) + DH(s(t))\xi$$

where $Df(s(t))$ and $DH(s(t))$ are the Jacobin matrices of f and H on the synchronization state $s(t)$, respectively. Then the equation (1-8) is rewritten as

$$\dot{\xi}_i(t) = Df(s(t))\xi_i + cDH(s(t))\sum_{j=1}^{N} a_{ij}\xi_j, \qquad (1\text{-}9)$$

Let $\xi = [\xi_1, ..., \xi_N]$, the equation (1-9) may be written as

$$\dot{\xi} = Df(s)\xi + cDH(s)\xi A^T$$

where $A^T = P\Lambda P^{-1}$ is Jordan decomposition of matrix A. Suppose the matrix Λ is diagonal, i.e. $\Lambda = diag\{\lambda_1, \cdots, \lambda_N\}$, where $\lambda_k (k=1,\cdots,N)$ are the eigenvalue of the matrix A. Let $\eta = [\eta_1, ..., \eta_N] = \xi P$ and $\eta_i (i=1,\cdots,N)$ is the eigenmode associated with the eigenvalue λ_i of A, then

$$\dot{\eta} = Df(s)\eta + cDH(s)\eta \Lambda \qquad (1\text{-}10)$$

$$\dot{\eta}_i = Df(s)\eta_i + c\lambda_i DH(s)\eta_i \qquad (1\text{-}11)$$

Since the coupling matrix has zero row-sum, there exists an eigenvalue being zero, with the corresponding eigenvector $\eta_1 = (1,1,\cdots,1)$. Therefore, the first eigenmode $\dot{\eta}_1 = Df(s)\eta_1$ corresponds to the perturbation parallel to the synchronization manifold. The other $N-1$ eigenmodes are transverse to synchronization manifold. Therefore, the equation (1-11) is equal to the following equation

$$\dot{\eta}_k = [Df(s) + c\lambda_k DH(s)]\eta_k, \quad k = 2,\ldots,N \qquad (1\text{-}12)$$

If the coupling matrix A is non symmetric, its eigenvalue may be complex. Suppose $c\lambda_k = \alpha + i\beta$, then the master stability equation (1-12) is rewritten as :

$$\dot{\varsigma} = [Df(s) + (\alpha + i\beta)DH(s)]\varsigma \qquad (1\text{-}13)$$

The largest Lyapunov exponent $L_{\max}$ of the equation (1-13) is the master stability function (MSF) of the network (1-6) [28, 32, 33, 36-38].

If the coupling matrix A is symmetric and un-weighted, its eigenvalues are as follows:

$$0 = \lambda_1 > \lambda_2 \geq \cdots \geq \lambda_N \qquad (1\text{-}14)$$

Clearly, it has only one eigenvalue being zero and all the other eigenvluses are less than zero. Denote $\alpha = c\lambda_l$, then the largest Lyapunov exponent $\lambda_{\max}(\alpha)$ is a function of α, which is the master stability function. The small η evolves as $\|\eta(t)\| \sim \exp(\lambda_{\max}(\alpha)t)$. Obviously, if $\lambda_{\max}(\alpha) < 0$, then $\|\eta(t)\| \to 0$, which implies that the corresponding eigenmode is stable[37,

39-41].The largest Lyapunov exponent linked to $\alpha = d\lambda_i$, the so-called MSF, fully determines the linear stability of the synchronous state. Moreover, the synchronous state, linked to $\lambda_1 = 0$, is stable if the remaining $N-1$ blocks, linked to $\lambda_i (i = 2,\cdots,N)$, have negative Lyapunov exponents [36]. Denote the region where the MSF $\Gamma(\alpha) < 0$ by $SR \subseteq R$. Depending on dynamical function f, inner coupling matrix h and synchronization state s, there are three possible types of networks [42]:

Type I network: $SR = (-\infty, \alpha_1)$, where α_1 is a negative finite integer. For this type of networks, the synchronization state is stable if

$$c\lambda_2 < \alpha_1 \tag{1-15}$$

or

$$c > |\alpha_1 / \lambda_2| \tag{1-16}$$

[43, 44]. This implies that synchronizability of Type I network can be characterized by the second largest eigenvlue λ_2 of the coupling matrix. Smaller λ_2 leads to better synchronizability.

Type II network: $SR = (\alpha_2, \alpha_3)$, where $-\infty < \alpha_2 < \alpha_3 < 0$. For this type of networks, the synchronization state is stable if $c\lambda_N > \alpha_2$ and $c\lambda_2 < \alpha_3$, which leads to

$$\lambda_N / \lambda_2 < \alpha_2 / \alpha_3 \tag{1-17}$$

[25, 33, 36]. For typical oscillators $\alpha_2 / \alpha_3 > 1$. This implies that the synchronizability of Type II network can be characterized by the eigenratio λ_N / λ_2 of the coupling matrix. Smaller λ_N / λ_2 leads to better synchronizability.

Type III network: $SR = \phi$. Synchronization in this type of networks can not be achieved with any coupling strength and coupling matrix.

Summarization of Synchronization in Complex Network

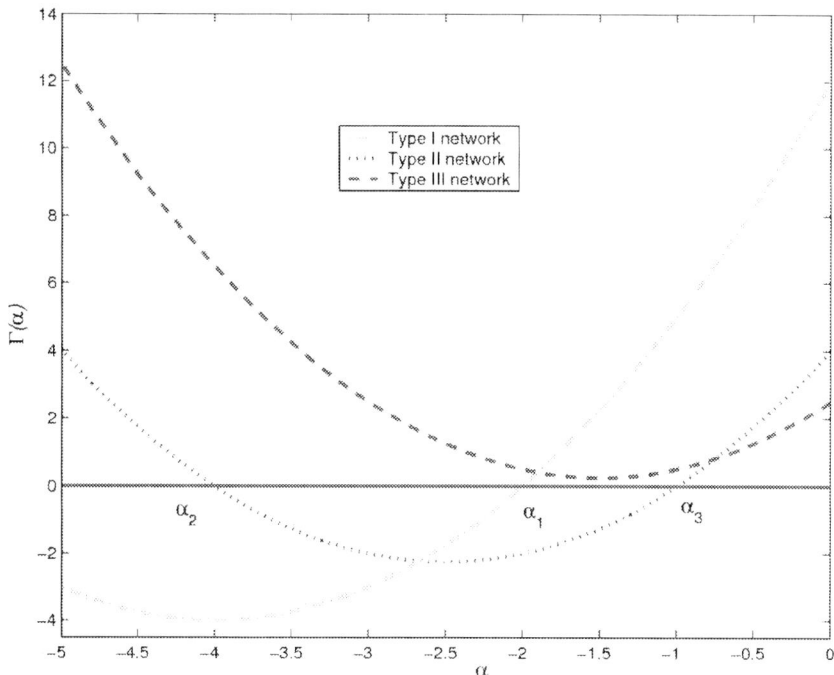

Figure 1.1. Illustration of the master stability functions of three network types(figure taken from [34], ©Elsevier).

Figure 1.1 illustrates the master stability functions of three network types. When the inner coupling matrix $h = diag(1,0,0)$ is a diagonal matrix, a typical of Type *I* network is the network of Chua's circuits, for which $\alpha_1 = -10$ [45]; and a Type *II* network is the network of Rossler oscillators, for which $\alpha_2 / \alpha_3 \approx 37.85$ [25].

Suppose the network is connected, the type I network must achieve synchronization for big enough coupling strength between nodes; while the type II network may reach synchronization for appreciate coupling strength scope. This implies that too strong or too weak coupling strength does not make the network achieve synchronization.

Beside the above three synchronization regions, Liu et al. found some network has different synchronization region such as $(-\infty, \alpha_1) \cup (\alpha_2, \alpha_3)$ [46]. By designing controllers, they adjust the network's synchronization region and improve the network synchronizability.

1.3.2. Synchronization of Un-weighted Networks

At first, synchronization of the un-weighted regular network with few nodes was investigated[33, 37, 47-49]. Since the discovery of many networks having the properties of scale-free and small-world, synchronization in these networks has been focused [25, 50].

Wang et al. found that type I small-world network has stronger synchronizability than the nearest network[45]. Barahona et al. investigated synchronization of the type II small-world, and found its synchronizability is also stronger than that of nearest network [25]. At first, the enhancement of the synchronization was contributed to the decrease of the average distance between nodes. However, Nishikawa et al. founded that when the increase of the degree homogeneity of their proposed two-layers' small world network, the synchronization of type II network is enhanced, while the average distance is decreased [51]. Hong et al. discovered that the synchronization of WS small world network is enhanced with the decrease of the degree homogeneity and the increase of the average path distance. This implies that the synchronization cannot be measured by the average path distance or the degree homogeneity. Duan et al found that the difference of the synchronization of two networks with the same betweenness is bigger [52]. Furthermore, if the synchronization region is finite, that is $S_2 = (\alpha_2, \alpha_3)$ in Section 1.3.1, the network synchronizability may be weakened or strengthened by adding edges; while for the unconnected complement network, the method of adding edges will never decrease the synchronizability. This implies it is important to investigate the synchronization of the network with the corresponding complement. Therefore, up to now, people have not found a network parameter to describe the whole network's synchronization.

Based on these two synchronization criteria, people investigated how to construct the un-weighted network with the optimal synchronizability. By adding synchronization preferential attachment in the network evolution, Fan et al. proposed synchronization optimal type I network model and synchronization preferential type I network, and found that the synchronization optimal type I network has similar multi-centers topology and high frangibility; while the synchronization preferential type I network high robustness, furthermore, its synchronizability is between the scale-free network and random graph [53, 54] According to the synchronization criterion II, Donetti et al. started a random network, used the method of analog annealing to rewire randomly, and obtained an synchronization optimal type II network [55]. The obtained network has extremely homogeneous topology, which means that some basic topology properties have narrow distributing scope. The detail introduction of synchronization in un-weighted networks may reference to[20, 56].

1.3.3. Synchronization of Weighted Networks

1.3.3.1. Adjust Node Degree

By setting the parameter $a_{ij} = l_{ij} / k_i^\beta$ in the equation (1-6), where l_{ij} is the entry of the Laplace matrix L, and its definition being

$$L_{ij} = \begin{cases} k_i & i = j \\ -1 & j \text{ is neighbor of node } i \\ 0 & otherwise \end{cases} \quad (1-18)$$

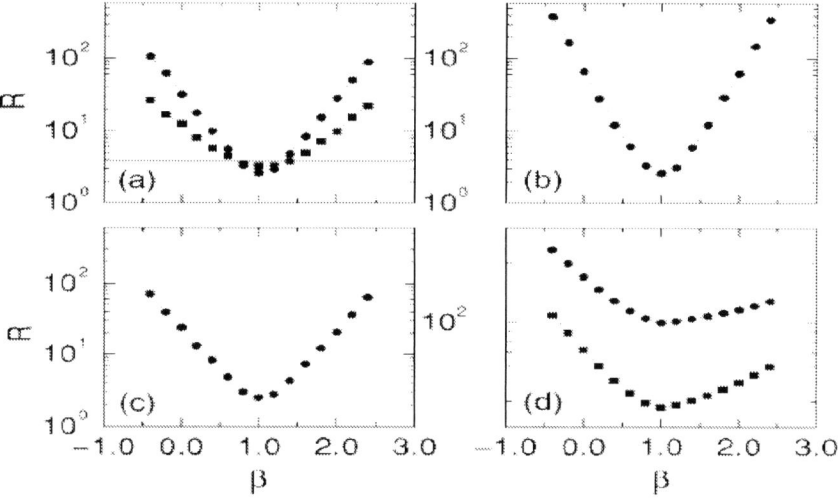

Figure 1.2. Eigenratio $R = \lambda_N / \lambda_2$ as function of β (a) random scale-free networks (b) networks with expected scale-free sequence (c) growing scale free networks (d) small world networks(figure taken from [31], ©American Physical Society).

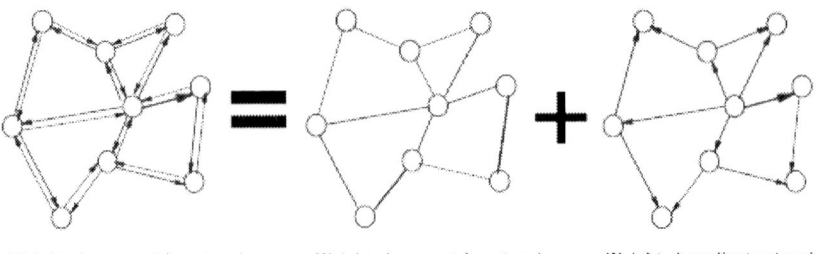

Weighted asymmetric network   Weighted symmetric network   Weighted gradient network

Figure 1.3. Schematic illustration of how a general weighted and asymmetrical network may be regarded as a superposition of a symmetrical and a directed (or gradient) network, both weighted (figure taken from [58], ©American Physical Society).

where k_i is the degree of node i, Motter et al. found that the type II network has strongest synchronizability for $\beta = 1$ in Figure 1.2[31]. Furthermore, the more heterogeneity of scale-free network, the better the method to improve the network synchronizability.

Korniss investigated synchronization of weighted networks with noise, where the coupling strength $a_{ij} = (k_i k_j)^{\beta}$. When the total coupling strength between nodes is fixed, it is also found that the network has strongest synchronizability for $\beta = -1$ [57].

The synchronization of gradient complex network is introduced in [58]. The coupling matrix of gradient network is often asymmetric, which means that $a_{ij} \neq a_{ji}$. Let $\Delta a_{ij} = a_{ij} - a_{ji}$, then $a_{ij} = (a_{ij} + a_{ji})/2 + \Delta a_{ij}/2$, where the first part is symmetric and the second part is directed. If the difference $\Delta a_{ij} > 0$, the coupling direction is from node j to node i. The decomposition is illustrated in Figure 1.3.

The gradient field exists often in temperature of chemical system and disposal velocity of neural networks[59, 60]. In the real world network, a random node is difficult to achieve the global information of the network. Then the gradient value of node i is due to the node's degree and its neighbors' degrees, that is

$$h_i = k_i^\beta \sum_{l \in v_i} k_l^\beta \qquad (1\text{-}19)$$

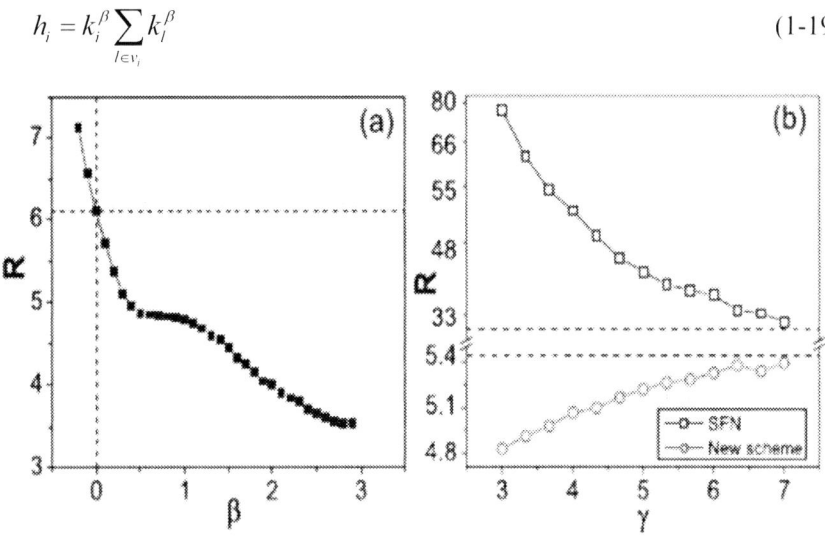

Figure 1.4. Ensemble of scale-free networks under the coupling scheme defined by (1-19). (a) The eigenratio R vs. β (b) R vs. γ (figure taken from [58] ,©American Physical Society).

Figure 1.3(a) shows the eigenratio R decreases with the increase of the value of β, this means that synchronization of the type II network is enhanced with the increase of the value of β. Especially, for $\beta > 0$, the coupling direction of the gradient network is from the node with bigger degree to the node with smaller degree. This implies that the node with bigger degree is more important than the node with smaller degree, which consists in reality network. As can be seen from Figure 1.4(b), for the un-weighted symmetric type II network, the eigenratio R decreases with the increase of the power-law index γ of degree distribution. Therefore, the synchronization of the type II un-weighted network is enhanced with the increase of degree heterogeneity. On the contrary, for the asymmetric gradient type II network, the eigenratio increases with the increase of the value of γ, which implies that synchronizability of type II gradient network is weakened with the increase of degree heterogeneity.

1.3.3.2. Adjust Coupling Direction and Edge Information

Inspired by the above Motter's work, Hwang et al. proposed a method of using age of node to enhance the synchronization of the network[61]. The earlier the node adding to the network, the bigger the age of the node; furthermore, the coupling strength between nodes with different ages is asymmetric. The coupling of this method is

$$G_{ij} = \begin{cases} -\theta_{ij} / \sum_{j \in \Lambda_i} \theta_{ij} & j \in \Lambda_i \\ 1 & j = i \\ 0 & otherwise \end{cases} \quad (1\text{-}20)$$

The definition of θ is

$$\theta_{ij} = \begin{cases} (1-\theta)/2 & i > j \\ (1+\theta)/2 & i < j, \end{cases} \quad (1\text{-}21)$$

where i, j are the node number, $i > j$ means that the node i adding to the network is earlier than node j, Λ_i is the set of the neighbors of node i. The parameter $-1 < \theta < 1$ determines that asymmetric grade of the coupling. Furthermore, for $\theta < 0$, the coupling effect from old node to new node is stronger; otherwise, the coupling effect from new node to old node is stronger. The case of $\theta = 0$ is the above Motter's work. Figure 1.5 shows that $\theta = -1$, the type II network has strongest synchronizability.

The above methods make the input reference signal being unitary, while this is not appropriate to many real world networks. Recently, Zhou and Chen investigated the case of the coupling being not unitary[62]. It is also found that when the coupling from old node to new node is dominant, the network has stronger synchronizability.

Chavez et al. proposed a method of using the edge information to improve the network synchronizability [63].The definition of their coupling method is:

$$G_{ij} = \begin{cases} -l_{ij}^{\alpha} / \sum_{j \in \Lambda_i} l_{ij}^{\alpha} & j \in \Lambda_i \\ 1 & j = i \\ 0 & otherwise \end{cases}, \qquad (1\text{-}22)$$

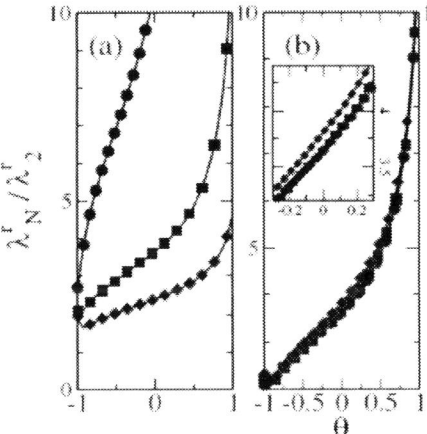

Figure 1.5. $\lambda_N^r / \lambda_2^r$ versus θ for (a) SF with different values of m ; (b) SF with different values of α (figure taken from [61] ,©American Physical Society).

where l_{ij} is the weight of the edge between node i and node j, α is a tunable parameter. As can be seen from (1-22), when $\alpha = 0$, this coupling method reduces to the Motter's coupling method for the case $\beta = 1$. For the scale free network with initial appeal and random network, it was found that the eigenratio λ_N / λ_2 is minimum for the case [63].

1.3.3.3. Optimal Synchronization of Weighted Networks

Similar to the un-weighted networks, the optimal synchronization of weighted networks has been a hot research topic for many scholars. Chavez et al. investigated the effect of associate coefficient on synchronization of networks, and found that the associate coefficient of optimal synchronization depends on the mechanism of weighted coupling[41]. When the network degree distribution is unchanged, they used the annealing method to adjust the associate coefficient of the network, and found that the smaller the associate coefficient, the weaker the synchronizability of type II network. Nishikawa and Motter found that: Beside the case that one node may connect to all the other nodes, a given type II network may reach optimal synchronization if the following three conditions are satisfied: (1) the network may be embedded an oriented spanning tree;(2) the network has no direction loop; and (3) the received coupling signal of each node is unitary. It is easy to obtain that the eigenratio of the coupling matrix is 1[64]. Combined the synchronization criterion II, the corresponding network has strongest synchronizability.

When the coupling matrix of the network is asymmetric and the eigenvalue of the coupling matrix is complex, the synchronization criterion is the MSF (1-13). However, it is difficult to calculate the Lyapunov exponent of the master stability equation corresponding to the complex eigenvalue. In order to solve this problem, a new synchronization criterion[65]. In this criterion, the second largest eigenvalue of symmetric matrix and the maximum of the line sum of asymmetric matrix are dominant.

1.3.3.4. Transition from Non-synchronization to Synchronization

As we know from Section 1.3.1, for a connected network, the eigenvalues of the coupling matrix are $0 = \lambda_1 < \lambda_2 \leq \cdots \leq \lambda_N$. The synchronization manifold is stable if all the MSF linked to master stability equation for the eigenvalues $\lambda_i (i = 2, \cdots, N)$ are negative [36]. In other words, when the coupling strength between nodes is bigger than α_1 or α_2 shown in Figure 1.1, the network is possible to achieve complete synchronization.

Figure 1-6 shows the master stability function MSF of type II network, linked to λ_2, is negative if $\alpha \in [\alpha_1 \lambda_2 / \lambda_i, \alpha_2 \lambda_2 / \lambda_i]$. Then according to scale relationship [25], the MSF, linked to $\lambda_i (i = 3, \cdots, N-1)$, is negative if $\alpha \in [\alpha_1 \lambda_2 / \lambda_i, \alpha_2 \lambda_2 / \lambda_i]$; and the MSF, linked to λ_N, is negative if $\alpha \in [\alpha_1 \lambda_2 / \lambda_N, \alpha_2 \lambda_2 / \lambda_N]$. Therefore, for $\alpha \in [\alpha_1, \alpha_2 \lambda_2 / \lambda_N]$, all the MSF, linked to all eigenvalues, is negative.

As can be seen from Figure 1.6, when α increases from 0 to $\alpha_1 \lambda_2 / \lambda_N$, the MSF, linked to λ_N, first begins to become negative. For further increase of α, the MSF, linked to $\lambda_i (i = 3, \cdots, N)$, becomes negative. When $\alpha \geq \alpha_1$, the MSF, linked to λ_2, finally becomes negative. For $\alpha \in [\alpha_1 \lambda_2 / \lambda_N, \alpha_1]$, in order to describe the transition of the network changes from non-synchronization to synchronization, a novel measure is proposed as follows [28]:

$$S = \frac{\alpha_1 r - \alpha_1 \lambda_2 / \lambda_N}{\alpha_1 - \alpha_1 \lambda_2 / \lambda_N} = \frac{r - \lambda_2 / \lambda_N}{1 - \lambda_2 / \lambda_N} \quad (1\text{-}23)$$

where $r\alpha_1 \in [\alpha_1 \lambda_2 / \lambda_N, \alpha_1]$ and $(\lambda_2 / \lambda_N) \leq r \leq 1$. The number of MSF distributed in the scope $[\alpha_1 \lambda_2 / \lambda_N, r\alpha_1]$ is assumed as q. For $r = \lambda_2 / \lambda_N$, $q = 1$, $S = 0$; for $r = 1$, $q = N$, $S = 1$. As shown in Figure 1.6, with the increase of r, the number of the negative MSF increases. When the value of r increases to 1, the whole MSFs became negative, which means the network may achieve synchronization. Therefore, the novel measure S may describe the synchronization transition of the network.

On the other hand, as can be seen from Figure 1.6, when α increases from $\alpha_2 \lambda_2 / \lambda_N$ to α_2, the network evolves from synchronization to non-synchronization. The symbol S in the equation (1-23) also can be used to describe this transition because:

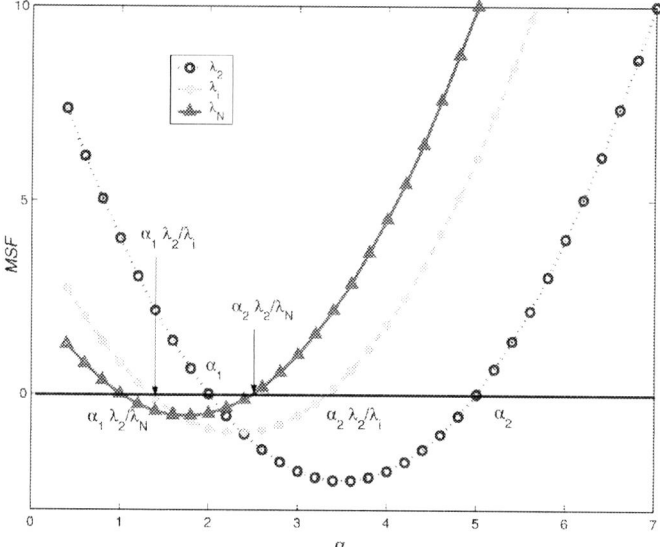

Figure 1.6. Illustration of the master stability functions of the corresponding eigenvalues $\lambda_i (i = 2, \cdots, N)$ (figure taken from [28], ©Elsevier).

$$S = \frac{\alpha_2 r - \alpha_2 \lambda_2 / \lambda_N}{\alpha_2 - \alpha_2 \lambda_2 / \lambda_N} = \frac{r - \lambda_2 / \lambda_N}{1 - \lambda_2 / \lambda_N} \qquad (1\text{-}24)$$

A BA scale-free network is constructed with $m = m_0 = 3$ and network size $N = 2000$. The parameter m is the number of the added new edges at each time step during the generation of the network. The detailed generation algorithm of BA scale-free network is introduced in [15]. Figure 1-7 (a) shows that the novel measure S vs. p of the scale-free network for different values of β with the network size $N = 2000$ and $m = m_0 = 3$, where $p = q / N$. As can be seen from Figure 1.7(a), S increases with the increase of p, which implies that transition of the network from non-synchronization to synchronization becomes faster with the increase of the number of MSF, which is negative.

For the same value of p, as shown in Figure 1.7(b), S increases with the increase of β from -1 to 0.4; however, as the value of β increases from 0.4 to 3, S decreases. Therefore, the transition of the network from non-synchronization to synchronization becomes faster is enhanced with the increase of the value of β; while as the value of β further increases, the transition of the network from non-synchronization to synchronization becomes slower. On the other hand, for the same value of β, the larger the value of p, the slower the transition of the network from non-synchronization to synchronization.

In addition, people investigated the synchronization of complex delay network[27, 66-72]; analyze the effect of network topology on synchronization[51, 73], and look for an effective control strategy and evolution mechanism to enhance the network synchronization[31, 66, 74-78].

The MSF method is only applied to the case of the network has fixed network topology. However, many real world network topology is time varying[79] [80] [27, 67, 68, 70, 81]. Belykh et al. proposed a connection graph stability method to measure the synchronization of time-varying network [27, 70, 81]. By calculating the total path lengths of the all edges of the network, a critical coupling strength of network synchronization is obtained. Wu investigated the synchronization of network with time delay and time-varying network topology, and obtain the condition of the network reaching global synchronization by constructing an appropriate Lyapunov function[82].

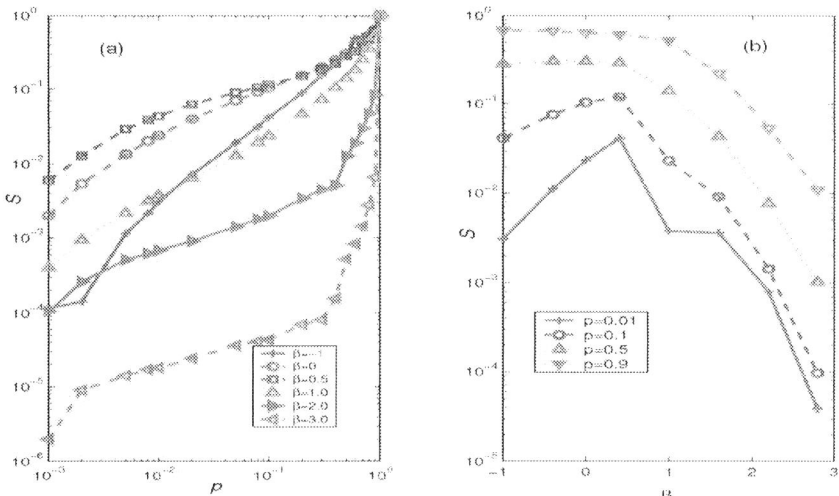

Figure 1.7. Transition of a scale-free network from non-synchronization to synchronization. (a) S vs. p, (b) S vs. β where $p = q/N$ (figures taken from [28] ©Elsevier).

Chapter 2

ADAPTIVE SYNCHRONIZATION OF COMPLEX NETWORKS

ABSTRACT

In this chapter, the adaptive method, which is commonly used in automation control field, is applied to make the complex network achieve synchronization. When the network topologies are unknown, an additional controller needs to be added to each node. Furthermore, the controller adaptively adjusts according to the error between the node's state and its destination state. On the other hand, when the network topologies are known, two adaptive approaches are proposed to make the network achieve synchronization. One is the global adaptive strategy, which implies the adaptive method uses the whole network's information; the other one is the local adaptive strategy, which uses only the information of the related node or edge.

2.1. INTRODUCTION

In most researches, the coupling strength is assumed to be identical for each edge in the network and constant in time. However, this assumption is often difficult to be satisfied since many real-world networks have evolving topology and adaptive coupling strength, which implies that the coupling strength change with time according to different environmental conditions. For example, for a group of robots, if one robot cannot see the front objects suddenly, the coupling strength between this robot and its neighbor must change in order to complete a difficult job[83]; more examples often may be found in biological networks and social insect colonies[84].

Recently, inspired by these potential and real applications, people proposed few adaptive strategies to investigate the synchronization of complex network [85-95]. We will introduce these recent adaptive approaches in brief.

2.2. ADAPTIVE SYNCHRONIZATION WITH UNKNOWN NETWORK TOPOLOGIES

Adaptive strategy was first proposed in the research of synchronization in complex network with uncertain network topologies [85]. The approach investigated a dynamical network with uncertain diffusively nonlinear coupling in the following form:

$$\dot{x}_i = f(x_i(t)) + g_i(x_1, x_2, \cdots, x_N) + v_i \qquad (2-1)$$

where g_i is smooth but unknown coupling functions, v_i is the control inputs and the other parameters are the same as those in (1-6). Obviously, when the network reach synchronization state $s(t)$ (1-7), the term $g_i(x_1, x_2, \cdots, x_N) + v_i = 0$, which means that

$$\dot{s}(t) = f(s(t)) \qquad (2-2)$$

Then the objects of design appreciate controllers v_i to make the network achieve synchronization, i.e.

$$\lim_{t \to \infty} \|x_i(t) - s(t)\| = 0$$

Subtracting (2-2) from (2-1) obtains the following error $e_i(t) = x_i(t) - s_i(t)$ system

$$\dot{e}_i(t) = f(x_i(t)) - f(s(t)) + g_i(x_1(t), x_2(t), \cdots, x_N(t)) - g_i(s(t), s(t), \cdots, s(t)) + v_i \qquad (2-3)$$

2.2. Local Synchronization

When the nodes' states are near to the synchronization state, linearize the system (2-3) on the synchronization manifold and obtain

$$\dot{e}_i(t) = Df(s(t)) + g_i(x_1(t), x_2(t), \cdots, x_N(t)) - g_i(s(t), s(t), \cdots, s(t)) + v_i \quad (2\text{-}4)$$

where $Df(s(t))$ is the Jacobin matrix on $s(t)$.

Assumption 2-1[85]. There exists positive constants α, β, and unknown but nonnegative constants γ_{ij}, $i, j = 1, 2, \cdots, N$, such that $\|Df(s(t))\| \leq \alpha$ and $g_i(x_1, x_2, \cdots, x_N) - g_i(s, s, \cdots, s) = B(t) E_i(x, s)$, satisfying $\|B(t)\| \leq \beta$, and

$$\|E_i(x, s)\| \leq \sum_{j=1}^{N} \gamma_{ij} \|e_j\| \quad (2\text{-}5)$$

Theorem 2-1[85]. Let assumption 2-1 hold and suppose there exists a matrix-valued function $P(t)$ which, for all t, is symmetrical and continuously differentiable, such that for some positive constants η, ρ, β, and δ,

$$\eta I \leq P(t) \leq \rho I, \quad (2\text{-}6)$$

$$(Df(s(t)) + B(t)k(t))^T P(t) + P(t)(Df(s(t)) + B(t)k(t)) + \dot{P}(t) \leq -\delta I \quad (2\text{-}7)$$

Then the following robust adaptive controllers, v_i and parameter estimation update laws, $\hat{\gamma}_i$ will robustly asymptotically stabilize the synchronization state:

$$v_i = B(t)(K(t)e_i - \frac{c_1 \hat{\gamma}_i B^T(t) P(t) e_i}{2}) \quad (2\text{-}8)$$

$$\dot{\hat{\gamma}}_i = c_1 \|B^T(t) P(t) e_i\|^2 \quad (2\text{-}9)$$

where $c_1 > \dfrac{N}{\delta}$, and $\hat{\gamma}_i$ are the estimates of the unknown parameters $\gamma_i = \sum_{j=1}^{N} \gamma_{ij}^2$ and initial values of $\hat{\gamma}_i(0) > 0$.

This theorem is proofed by constructing a Lyapunov function. The isolated node's dynamics is selected to be the chaotic Chen system as follows:

$$\begin{cases} \dot{x}_1 = 35(x_2 - x_1) \\ \dot{x}_2 = -7x_1 - x_1 x_3 + 28x_2 \\ \dot{x}_3 = x_1 x_2 - 3x_3 \end{cases} \quad (2\text{-}10)$$

When the matrix of the controllers in (2-8) and (2-9) are assigned

$$B(t) = \begin{pmatrix} 1 & 0 \\ 0 & 1 \\ 0 & 0 \end{pmatrix}, \quad P(t) = \begin{pmatrix} 1 & 0 & 0 \\ 0 & 1 & 0 \\ 0 & 0 & 1 \end{pmatrix}, \quad K(t) = \begin{pmatrix} 0 & s_3 - 28 & -s_2 \\ 0 & -k & 0 \end{pmatrix}$$

(2-11)

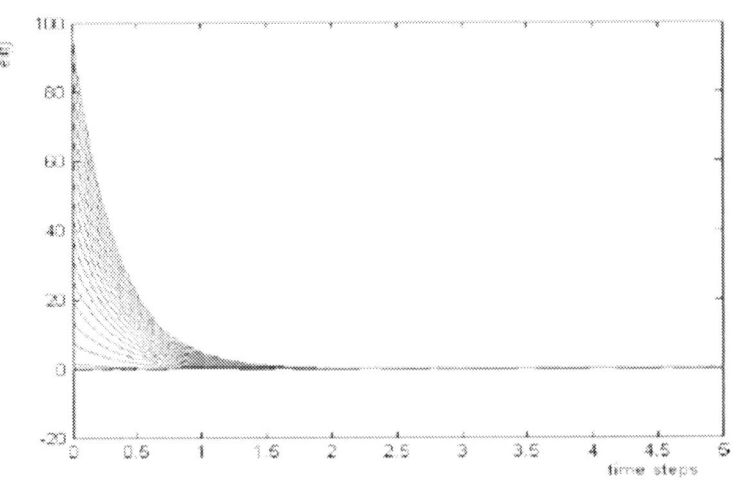

Figure 2.1. Synchronization errors $e_i(t)$ between the state of Chen system and the state of uncertain network with unknown nonlinear coupling using the designed robust adaptive controllers(figure taken from [85], ©Elsevier).

where $s = (s_1, s_2, s_3)^T$. Figure 2.1 shows that 50 Chen systems reach the synchronization state.

As can be seen from Theorem 2-1, when the network size is big, it is difficult to obtain the matrix $P(t), K(t)$ in the corresponding adaptive controllers. Zhou et al. proposed a simple adaptive controllers[96].

Theorem 2-2[96]. Let Assumption 2-1 hold, the synchronization state $s(t)$ of uncertain dynamical network (2-1) is locally asymptotically stable under the adaptive controllers

$$v_i = -d_i e_i \tag{2-12}$$

and updating laws

$$\dot{d}_i = k_i e_i^T e_i \tag{2-13}$$

where k_i are positive constant.

Proof: Construct a Lyapunov candidate as follows:

$$V = \frac{\sum_{i=1}^{N} e_i^T e_i}{2} + \frac{1}{2}\sum_{i=1}^{N} \frac{(d_i - \hat{d}_i)^2}{k_i} \tag{2-14}$$

where $\hat{d}_i$ are positive constants. Thus, the derivative of V on time t obtains

$$\begin{aligned}
\frac{dV}{dt} &= \frac{\sum_{i=1}^{N}(\dot{e}_i^T e_i + e_i^T \dot{e}_i)}{2} + \sum_{i=1}^{N} \frac{(d_i - \hat{d}_i)\dot{d}_i}{k_i} \\
&= \sum_{i=1}^{N} e_i^T \left(\frac{Df(s(t)) + D^T f(s(t))}{2} - d_i I\right) e_i + \\
&\quad \sum_{i=1}^{N} e_i^T (g_i(x_1(t), x_2(t), \cdots, x_N(t)) - g_i(s(t), s(t), \cdots, s(t))) + \sum_{i=1}^{N}(d_i - \hat{d}_i) e_i^T e_i \\
&\leq \sum_{i=1}^{N} e_i^T \left(\frac{Df(s(t)) + D^T f(s(t))}{2} - \hat{d}_i I\right) e_i + \sum_{i=1}^{N} e_i^T \sum_{j=1}^{N} \gamma_{ij} \|e_j\|
\end{aligned}$$

$$\leq \sum_{i=1}^{N}(\alpha - \hat{d}_i)e_i^T e_i + \sum_{i=1}^{N}\sum_{j=1}^{N}\gamma_{ij}e_i^T \|e_j\|$$

$$= e^T(\Gamma + \Lambda)e$$

where $\Gamma = (\gamma_{ij})_{N \times N}$ and Λ is the diagonal matrix with diagonal entries $\alpha - \hat{d}_i$. Select the value of α and γ_{ij} to make the matrix $\Gamma + \Lambda$ be negative. Thus the network locally stabilizes on the synchronization state under the adaptive controllers (2-12) and updating laws (2-14).

The proof is completed.

2.2.2. Global Synchronization

The results of Theorem 2-1 and 2-2 are based on the linear method around the synchronization state. However, in many real world networks, the nodes' states are randomly distributed. Rewrite the dynamical network (2-1) in a general system as

$$\dot{x}_i = Ax_i(t) + f(x_i(t)) + g_i(x_1, x_2, \cdots, x_N) + v_i \qquad (2\text{-}15)$$

where $A \in R^{n \times n}$ is a constant matrix, the definitions of the other parameters are the same as those in (2-1).

Substrate (2-2) from (2-15) and obtain the corresponding error system

$$\dot{e}_i(t) = Ae_i(t) + f(x_i(t)) - f(s(t)) + g_i(x_1(t), x_2(t), \cdots, x_N(t)) - g_i(s(t), s(t), \cdots, s(t)) + v_i$$

$$(2\text{-}16)$$

Assumption 2-3[96]. There exists positive constants β, and unknown but nonnegative constants μ_i and γ_{ij}, $i, j = 1, 2, \cdots, N$, such that

$$\|f(x_i(t)) - f(s(t))\| \leq \mu_i e_i,$$

$$\|g_i(x_1(t), x_2(t), \cdots, x_N(t)) - g_i(s(t), s(t), \cdots, s(t))\| \leq \sum_{j=1}^{N}\gamma_{ij}\|e_j\| \qquad (2\text{-}17)$$

Theorem 2-3[96]. Let Assumption 2-3 hold, the synchronization state $s(t)$ of uncertain dynamical network (2-15) is globally asymptotically stable under the adaptive controllers

$$v_i = -d_i e_i \tag{2-18}$$

and updating laws

$$\dot{d}_i = k_i e_i^T e_i \tag{2-19}$$

where k_i are positive constant.

The proof may reference to the proof of Theorem 2-2.

Consider a dynamical network made of 50 Lorenz systems. The isolated node dynamics is described by

$$\begin{pmatrix} \dot{x}_{i1} \\ \dot{x}_{i2} \\ \dot{x}_{i3} \end{pmatrix} = A \begin{pmatrix} x_{i1} \\ x_{i2} \\ x_{i3} \end{pmatrix} + \begin{pmatrix} 0 \\ -x_{i1} x_{i3} \\ x_{i1} x_{i2} \end{pmatrix}$$

where

$$A = \begin{pmatrix} -a & a & 0 \\ c & -1 & 0 \\ 0 & 0 & b \end{pmatrix}$$

Here $a=10, b=8/3, c=28$, and $i=1,2,\cdots,50$. The networked system is defined as follows:

$$\begin{pmatrix} \dot{x}_{i1} \\ \dot{x}_{i2} \\ \dot{x}_{i3} \end{pmatrix} = A \begin{pmatrix} x_{i1} \\ x_{i2} \\ x_{i3} \end{pmatrix} + \begin{pmatrix} 0 \\ -x_{i1} x_{i2} \\ x_{i1} x_{i2} \end{pmatrix} + \begin{pmatrix} f_1(x_i) - 2f_1(x_{i+1}) + f_1(x_{i+2}) \\ 0 \\ f_2(x_i) - 2f_2(x_{i+1}) + f_2(x_{i+2}) \end{pmatrix} + d_i e_i \tag{2-20}$$

and the update laws

$$\dot{d}_i = k_i e_i^T e_i, \tag{2-21}$$

$f_1(x_i) = a(x_{i2} - x_{i1}), f_2(x_i) = x_{i1}x_{i2} - bx_{i3}, x_{51} \equiv x_1, x_{52} \equiv x_2$. It is easy to obtain that

$$\|f(x_i(t)) - f(s(t))\| = \left\| \begin{pmatrix} 0 \\ -x_{i1}x_{i3} + s_1 s_3 \\ x_{i1}x_{i2} - s_1 s_2 \end{pmatrix} \right\|$$

$$= \left\| \begin{pmatrix} 0 \\ -x_{i1}e_{i1} - s_1 e_{i3} \\ x_{i2}e_{i1} + s_1 e_{i2} \end{pmatrix} \right\| = \sqrt{(-x_{i1}e_{i1} - s_1 e_{i3})^2 + (x_{i2}e_{i1} + s_1 e_{i2})^2} \tag{2-22}$$

As known from [97], the Lorenz system is bounded, which means that there exists a constant M satisfying $|x_{ij}|, |s_j| < M$ for $i = 1, 2, \cdots, 50, j = 1, 2, 3$. Therefore the (2-23) becomes that

$$\|f(x_i(t)) - f(s(t))\| \le 2M \|e_i\| \tag{2-23}$$

Similarly, one has

$$\|g_i(x_1(t), x_2(t), \cdots, x_N(t)) - g_i(s(t), s(t), \cdots, s(t))\|$$
$$= \left\| \begin{pmatrix} f_1(x_i) - 2f_1(x_{i+1}) + f_1(x_{i+2}) \\ 0 \\ f_2(x_i) - 2f_2(x_{i+1}) + f_2(x_{i+2}) \end{pmatrix} \right\| \le 3\sqrt{2(a^2 + M^2)}(\|e_i\| + \|e_{i+1}\| + \|e_{i+2}\|)$$

$$\tag{2-24}$$

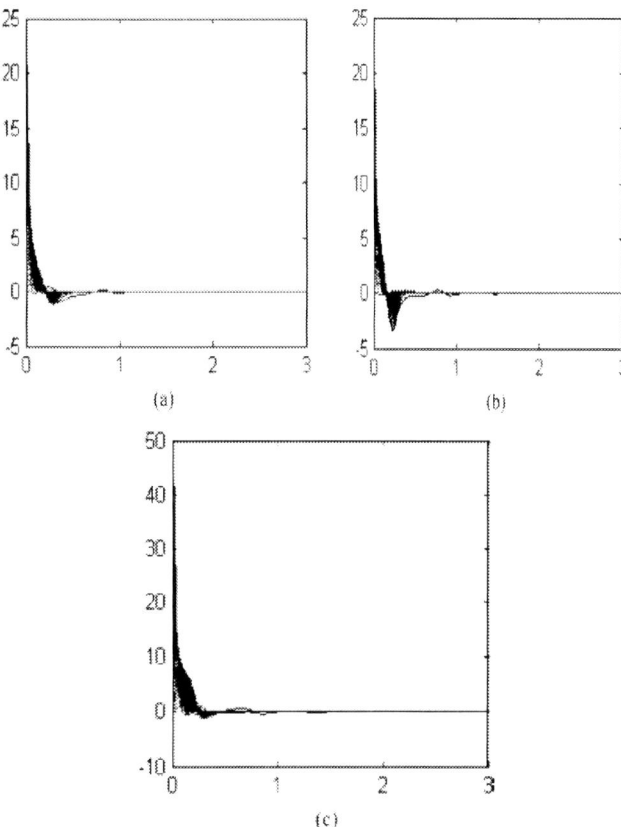

Figure 2.2. Synchronization errors of network (2-20) and (2-21) with the increase of time t. (a) $e_{i1}(t)$ (b) $e_{i2}(t)$ (c) $e_{i3}(t)$ (figure taken from [96], © IEEE).

From (2-23) and (2-24), it is easy to find the assumption 2-3 satisfies, then the network will globally reach synchronization according to Theorem 2-3. Figure 2.2 shows that synchronization error may reach zero for $k_i = 1$, $d_i(0) = 1$, $x_i(0) = (4 + 0.5i, 5 + 0.5i, 6 + 0.5i)$ with $i = 1, 2, \cdots, 50$, and $s(0) = (4, 5, 6)$. This means the network globally synchronization.

2.3 ADAPTIVE SYNCHRONIZATION WITH KNOWN NETWORK TOPOLOGIES

When the network topologies are known, several adaptive strategies are proposed to enhance the synchronization in complex network. Based on the scope of the used information, these adaptive strategies may be classified into two kinds: global information method and local information method.

2.3.1. Global Information

De Lellis et al. introduced adaptive strategy based on the global information of the network [93]. Similar to the equation (1-6) in Chapter 1, the node state is in the form

$$\dot{x}_i(t) = f(x_i(t)) + \sum_{j=1}^{N} g_{ij}[H(x_j(t)) - H(x_i(t))], \qquad (2\text{-}25)$$

where $G = (g_{ij})$ is the weighted coupling matrix with the entries $g_{ij} = a_{ij}w$. Here the adjacency matrix $A = (a_{ij})_{N \times N}$ is binary. If there exists a connection between node i and node j, then $a_{ij} = 1$; otherwise, $a_{ij} = 0$. w is the coupling strength between nodes if they are connected. Especially, $w = 1$ for all edges in the unweighted network. The other parameters' definitions are the same as those in (1-6).

Their proposed adaptive law is as follows:

$$\dot{w}(t) = \Phi(x_1, x_2, \cdots, x_N) \qquad (2\text{-}26)$$

where the function $\phi(\cdot)$ is relevant to all the nodes' states in the network. For example, the function $\phi(\cdot)$ may be selected in the form:

$$\dot{w}(t) = \frac{\mu}{N}\sum_{j=1}^{N}\left\|H(x_j(t)) - \frac{\sum_{k=1}^{N}H(x_k(t))}{N}\right\|$$ (2-27)

where μ is the adaptive gain.

2.3.2. Local Information

Recently, Zhou and Kurth first proposed an adaptive strategy based on local information to investigate synchronization of complex networks in (2-25) [86]. Here $g_{ij} = a_{ij}w_{ij}$ and w_{ij} is the coupling strength from node j to node i. Their adaptive law is

$$\dot{w}_{ij} = \gamma\Delta_i / (1+\Delta_i)$$ (2-28)

where

$$\Delta_i = \left|H(x_i) - \frac{\sum_j a_{ij}H(x_j)}{\sum_j a_{ij}}\right|$$ (2-29)

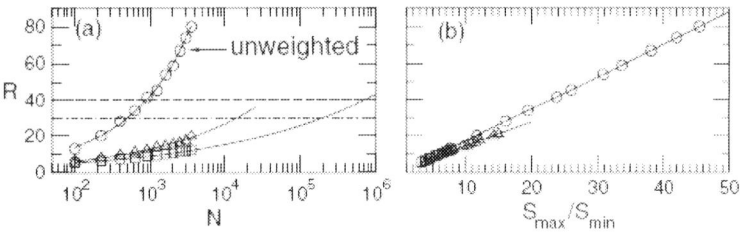

Figure 2.3. The eigenratio $R = \lambda_N / \lambda_2$ as a function of the network size N (a) and S_{max} / S_{min} (b), average over 20 realizations of the networks. The networks may be synchronized if $R < R_c$ in (a): Rossler oscillators, $R_c = 40$ (dashed line), food web model $R_c = 29$ (dashed-dotted line) (figure taken from [86] ©American Physical Society).

and $\gamma > 0$ is adaptive gain.

Furthermore, the strength to node i from all its neighbors increases uniformly among these edges between them, which means that $v_i = w_{ij}$ for any j. Then the difference Δ_i of node i may be reduced.

Figure 2.3 shows that synchronization of type II scale-free network. The network is generated from 5 fully connected nodes, at each time a new node is added to the network with 5 new edges between the new node and the old nodes in the network [15]. The adaptive gain is $\gamma = 0.002$ with $H(x) = (x,0,0)$ for the Rossler oscillators (□), and $H(x) = (0, y, z)$ for the food web model (△). The intensity S_i of node i is defined as $S_i = v_i \sum_j a_{ij}$. The maximum of all nodes' intensities is defined as $S_{max} = \max(S_i)$, while the minimum intensity is $S_{min} = \min(S_i)$. As can be seen from Figure 2.3(a), the eigenratio R is obviously reduced by introducing the adaptive strategy. According to the second synchronization criterion, the synchronizability is enhanced in the adaptive network. Figure 2.3(b) shows that the change of eigenratio is consistent with the change of the ratio S_{max} / S_{min}. Therefore, the ratio S_{max} / S_{min} may be used to measure the synchronizability of a complex network.

Based on the Zhou's adaptive approach, De Lellis et al. proposed two simpler adaptive strategy only based on the local information to investigate the synchronization of complex networks: edge-based strategy and vertex-based strategy; the former uses local adaptive coupling strength at each node, the latter depends only on the local information of the corresponding edge and nodes [92-94].

2.3.2. Vertex-based Strategy

The adaptive law for the coupling strength g_{ij} in (2-25) is selected as

$$\dot{g}_{ij}(t) = \mu \left\| \sum_{k \in \varepsilon_i} [(H(x_k) - H(x_i)] \right\| \tag{2-30}$$

where ε_i is the set of the neighbors of node i. Obviously, for any node of ε_i, the adaptive law is same, which implies $\dot{g}_{ij}(t) = \dot{g}_i(t)$. Therefore, each node has a different coupling strength.

2.3.2.2. Edge-based Strategy

Each edge has a different coupling strength and its adaptive law

$$\dot{g}_{ij}(t) = \alpha \|H(x_k) - H(x_i)\| \tag{2-31}$$

where α is the adaptive gain.

The Laplace matrix L describes the network topology and its entry is l_{ij}. If there is a connection from node i to node j, then $l_{ij} = -1$; otherwise, $l_{ij} = 0$. Furthermore, $l_{ii} = -\sum_{j, j \neq i} l_{ij}$.

If the Laplace matrix L is symmetric and un-weighted, its eigenvalues are as follows:

$$0 = \lambda_1 < \lambda_2 \leq \cdots \leq \lambda_N$$

The normalized eigenvector corresponding to the zero eigenvalue of the Laplace matrix L is written as $\xi = (\xi_1, \xi_2, \cdots, \xi_N)^T$. Let the matrix $\Xi = diag\{\xi_1, \xi_2, \cdots, \xi_N\}$ and the matrix $U = \Xi - \xi \xi^T$. The equation of the network (2-25) may be rewritten as follows:

$$\dot{X} = F(X) + (G \otimes I_n)X$$

where $X = (x_1^T, \cdots, x_N^T)^T, F(X) = (f(x_1)^T, \cdots, f(x_N)^T)^T$.

Theorem 2-4. If there exists a positive definite matrix $P = diag\{p_1, \cdots, p_n\}$, an arbitrary diagonal matrix $\Delta = diag\{\Delta_1, \cdots, \Delta_n\}$, and a constant $\bar{\omega} > 0$, the inequation

$$(x-y)^T P(f(x)-f(y)+\Delta y - \Delta x) \leq -\bar{\omega}(x-y)^T(x-y), \forall x, y \in R^n, \tag{2-32}$$

satisfies and the matrix $(U \otimes I_n)(I_n \otimes \Delta) + (U \otimes I_n)(G \otimes I_n)$ is negative semi definite, where $\otimes$ is the Kronecker product, then the edge-based strategy guarantees synchronization [92].

Proof: Consider a candidate Lyapunov function as follows

$$V = \frac{1}{2}\eta X^T(t)(U \otimes I_n)X(t) + \frac{1}{2\alpha}\sum_\varepsilon (c_{ij} - g_{ij}(t))^2 \tag{2-33}$$

where η is a positive scalar and c_{ij} is an arbitrary scalar associated to each edge, ε is the set of the edges in the network.

Derivate V on t, it is obtained that

$$\begin{aligned}\dot{V} &= \eta X^T(t)(U \otimes I_n)\dot{X}(t) - \frac{1}{\alpha}\sum_\varepsilon (c_{ij} - g_{ij}(t))\dot{g}_{ij}(t) \\ &= \eta X^T(t)(U \otimes I_n)[F(X) + G \otimes I_n X] - \sum_\varepsilon (c_{ij} - g_{ij}(t))\|x_j - x_i\|\end{aligned} \tag{2-34}$$

Adding and subtracting $X^T(U \otimes I_n)(I_n \otimes \Delta)X$, obtains

$$\begin{aligned}\dot{V} &= \eta X^T(t)(U \otimes I_n)[F(X) - (I_n \otimes \Delta)X] + G \otimes I_n X] \\ &+ \eta X^T(t)(U \otimes I_n)(I_n \otimes \Delta)X - \sum_\varepsilon (c_{ij} - g_{ij}(t))\|x_j - x_i\|X^T(U \otimes I_n)\end{aligned} \tag{2-35}$$

Since the inequation (2-31) is satisfied, then $X^T(t)(U \otimes I_n)[F(X) - (I_n \otimes \Delta)X] \leq -\bar{\omega}X^T(t)(U \otimes I_n)X$. Therefore, (2-33) may be written as follows:

$$\dot{V} \leq -\varpi\eta X^T(t)(U \otimes I_n)X + \eta X^T(t)[(U \otimes I_n)(I_n \otimes \Delta) + (U \otimes I_n)(G \otimes I_n)]X$$
$$- \sum_{\varepsilon}(c_{ij} - g_{ij}(t))\|x_j - x_i\| X^T(U \otimes I_n)$$

(2-36)

As known from (2-31) and the matrix $(U \otimes I_n)(I_n \otimes \Delta) + (U \otimes I_n)(G \otimes I_n)$ is negative semi definite, then the first two parts of the left side of (2-34) are negative. Furthermore, for any edge $(i,j) \in \varepsilon$, there exists the value c_{ij} is bigger than the corresponding edge strength g_{ij}, it is obtained that $\dot{V} < 0$.

The proof is completed.

Theorem 2-5[92] If there exist a positive definite matrix $P = diag\{p_1, \cdots, p_n\}$, an arbitrary diagonal matrix $\Delta = diag\{\Delta_1, \cdots, \Delta_n\}$, and a constant $\bar{\omega} > 0$, the in-equation (2-31) satisfies and the matrix $(L \otimes I_n)(I_n \otimes \Delta) + (L \otimes I_n)(G \otimes I_n)$ is negative semi definite, then the vertex-based strategy guarantees synchronization.

In the above two theorems, the in-equation (2-32) is difficult to verify. As known from Ref. [98], many chaotic systems satisfies (2-32). Furthermore, when the matrix $\Delta = 0$, then obtain a simpler condition to be verified.

Corollary 2-1[92]: If there exist a positive definite matrix $P = diag\{p_1, \cdots, p_n\}$, and a constant $\bar{\omega} > 0$, the in-equation

$$(x-y)^T P(f(x)-f(y)) \leq -\bar{\omega}(x-y)^T(x-y), \forall x, y \in R^n,$$

(2-37)

(2-31) satisfies, both the edge-based and the vertex-based strategies guarantee synchronization.

The Chua's circuit is considered as each isolated node's dynamics in the complex dynamical network:

$$\begin{cases} \dot{y}_1 = u[y_2 - h(y_1)] \\ \dot{y}_2 = y_1 - y_2 + y_3 \\ \dot{y}_3 = -vy_2 \end{cases}$$

(2-38)

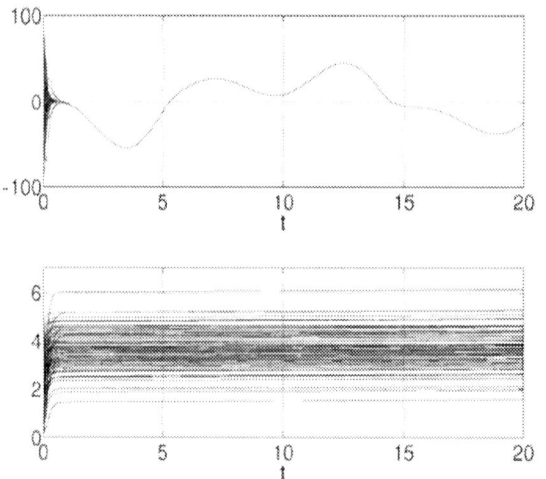

Figure 2.4. Synchronization of a scale-free network of 100 Chua's Circuits using the vertex-based adaptive method: evolution of $x_i (i = 1, 2, \cdots, 100)$ (top) and g_i (bottom) (figure taken from [92], ©American Institute of Physics).

where $h(x) = -0.07x - 0.785[|x+1| - |x-1|]$, $u = 59/12$, and $v = 295/81$ [92]

A BA scale-free network is constructed with $m = m_0 = 5$ and $N = 100$ [15]. The initial states of the Chua's oscillators are randomly distributed in a normal distribution with mean equal to 0 and standard deviation equal to 40. The adaptive gains in these two adaptive strategies are 0.1. As known from Ref. [98], the Chua's oscillator satisfies the in-equation (2-31). This means that the network may reach synchronization for both adaptive strategies according to the Corollary 2-1. Figure 2.4 (a) shows the nodes reaches synchronization for $t > 2s$ when the vertex-based adaptive method is adopted. As can be seen from Figure 2.4(b), the coupling strength of each node rapidly increases for $t < 2s$. Since the left side of (2-28) is big for $t < 2s$. For $t > 2s$, the network reaches synchronization, which leads to the left side of (2-28) is zero. This means that the coupling strength of each node do not change any more. Similar phenomenon also may be found in Figure 2.5 with the edge-based adaptive method.

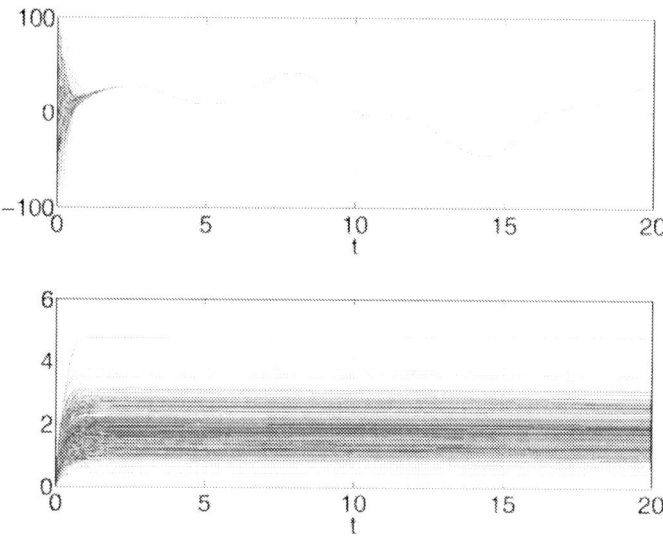

Figure 2.5. Synchronization of a scale-free network of 100 Chua's circuits using the edge-based adaptive method: evolution of $x_i (i = 1, 2, \cdots, 100)$ (top) and g_{ij} (bottom) (figure taken from [92], ©American Institute of Physics).

Chapter 3

CLUSTER SYNCHRONIZATION IN COMPLEX NETWORKS

ABSTRACT

In many technological, social and biological networks, the network are made up of several subgroups. When the nodes in the same subgroup achieve one state, the nodes in different subgroups achieve different states, the network is said to achieve cluster synchronization. At first, the methods by finding an appropriate coupling strength matrix are presented to make the network achieve cluster synchronization. However, this method is difficult to be applied especially when the network has lots of nodes and complex connections between nodes. Then, by adding a simple controller to each node, the network with random divided subgroups may achieve cluster synchronization irrespective of the coupling strength no more.

3.1. INTRODUCTION

The synchronization investigated in Chapter 1 and 2 is complete synchronization; this means that when the network reach synchronization, all the nodes reaches the same synchronization state. However, many technological, social and biological networks divide naturally into subgroups, and nodes in the same subgroup often make same kind of function [99-106]. For example, when a crowd of people choose to accept or oppose an opinion according to one's preference, two subgroups are often formed: one subgroup is made up of those who accept the opinion and the other one is made up of those who oppose the opinion[99]. Subgroups in neuronal networks and biochemical networks may also represent functional subgroups [106] and pages on relevant topics may make up of subgroups on the web[104].

Cluster synchronization is proposed to investigate these problems, which means that the whole nodes in the network are split into several subgroups, nodes in the same subgroup achieve the same synchronization state, while nodes in different subgroups achieve different synchronization state [101, 107-122]. In this chapter, we will briefly introduce these recent researches.

3.2. SELECT APPRECIATE COUPLING MATRIX

At first, by selecting a particular coupling structure, Belykh et al. investigated the cluster synchronization of complex networks [107]. Consider N coupled nodes described by the following system:

$$\dot{X} = F(X) + (\varepsilon G \otimes H)X \qquad (3\text{-}1)$$

where $X = (X_1, X_2, \cdots, X_N)^T$ is the set of dynamical variables, and $F(X) = (F(X_1), F(X_2), \cdots, F(X_N))^T$. The matrix ε describes the nonlocal type of coupling as well as the coupling strength between nodes. The investigated matrix G is diffusion or nearest-neighbor coupling matrix

$$G = \begin{pmatrix} -1 & 1 & 0 & \cdots & 0 \\ 1 & -2 & 1 & \cdots & 0 \\ \vdots & \vdots & \vdots & \vdots & \vdots \\ 0 & \cdots & 0 & 1 & -1 \end{pmatrix}$$

This type of nonlocal coupling is that each element of the array in principle is coupled with each other through its local diffusive interaction with the nearest neighbors. The matrix $A = \varepsilon G$ is arbitrary with the sum of all elements of each line being zero, which means that there exists complete synchronization manifold. Especially, the matrix $\varepsilon = \varepsilon_c = \varepsilon I_N$, the system becomes a system of locally coupled.

When the locally coupled networks with the matrix ε_c do not happen cluster synchronization and only complete synchronization occurs, Belykh et al. found that there exists the matrix ε of non-locally coupled networks happens cluster synchronization, while the complete synchronization is unstable.

Let the matrix $E_m = I_m \otimes I_n$ and $\bar{E}_m = \bar{I}_m \otimes I_n$, where I_n and I_m are unit matrices, and $\bar{I}_m$ is a $n \times n$ matrix whose nonzero entries are 1 and lie in the secondary diagonal. $C_e = (E_m \bar{E}_m)^T$ for an even number of the network size $N = 2m$,

$$C_0 = \begin{pmatrix} E_{m-1} & 0 \\ 0 & I_n \\ \bar{E}_{m-1} & 0 \end{pmatrix}$$

for odd $N = 2n-1$, $C_a = (E_m, \bar{E}_m, E_m, \cdots)^T$, where E_m alternating with $\bar{E}_m$ is repeated r times, for $N = rm$.

Proposition 3-1[107]: Let the matrix $\underline{\varepsilon C_s}$ represent the first m lines of the matrix εC_s and let

$$C_s(\underline{\varepsilon C_s}) = \varepsilon C_s \tag{3-2}$$

where s represents o, e and a. Then the network (3-1) with the global coupling matrix ε has m-cluster synchronization manifolds: $M^c(2m, m) = \{X = C_e U\}$ for even $N = 2m$, $M^c(2m-1, m) = \{X = C_o U\}$ for odd $N = 2m-1$, and $M^c(rm, m) = \{X = C_a U\}$ for $N = rm$.

When the network size $N=5$, for the local diffusive coupling, the network (3-1) has only two synchronization manifolds: one is $M(5,1)=\{X_1=X_2=\cdots=X_5\}$ that corresponds to complete synchronization of network and the other one $M^c\{5,3\}=\{X_1=X_5, X_2=X_4\}$ that corresponds to three clusters. According to the proposition 3-1, the cluster synchronization manifold $M^c\{5,3\}$ exists when

$$\varepsilon = \begin{pmatrix} 1 & 9 & 4 & 3 & 5 \\ 9 & 6 & 5 & 2 & 0 \\ 4 & 1 & 9 & 7 & 3 \\ 1 & 7 & 5 & 1 & 8 \\ 4 & 7 & 4 & 5 & 2 \end{pmatrix}, C_o = \begin{pmatrix} 1 & 0 & 0 \\ 0 & 1 & 0 \\ 0 & 0 & 1 \\ 0 & 1 & 0 \\ 1 & 0 & 0 \end{pmatrix} \tag{3-3}$$

Obviously, the matrix ε is complex. Belykh et al. considered the matrix ε having only nonzero elements in the principal and secondary diagonals (all other elements are zeros) [107]. Figure 3.1 shows that 5 Sherman models are classified into three clusters. At the same time, the system is

$$\begin{cases} \dot{X}_1 = F(X_1) - 4H(X_4 - X_5) \\ \dot{X}_2 = F(X_2) + 4H(X_3 - 2X_2 + X_1) \\ \dot{X}_3 = F(X_3) + 4H(X_4 - 2X_3 + X_2) \\ \dot{X}_4 = F(X_4) + 4H(X_5 - 2X_4 + X_3) \\ \dot{X}_5 = F(X_5) - 4H(X_2 - X_1) \end{cases}$$

where the inner coupling matrix is defined as

$$H = \begin{pmatrix} 1 & 0 & 0 \\ 0 & 0 & 0 \\ 0 & 0 & 0 \end{pmatrix}.$$

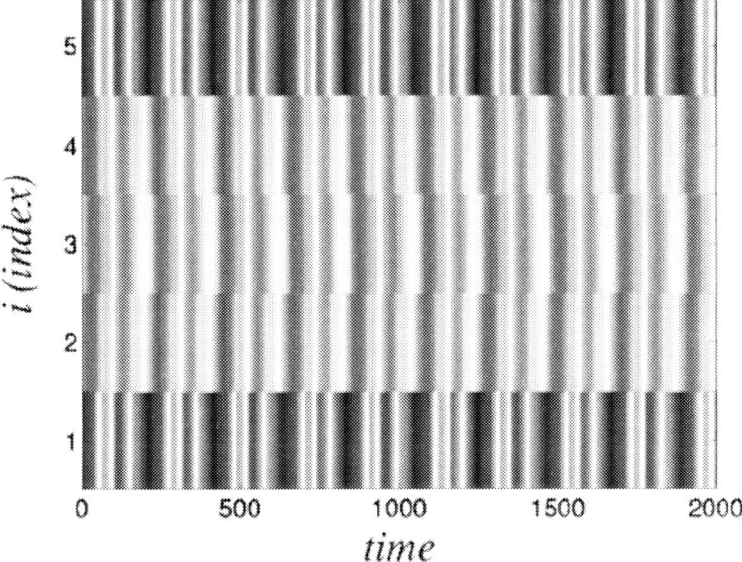

Figure 3.1. Cluster synchronization of 5 nonlocally coupled Sherman nodes (figure taken from [107] ©American Physical Society).

Furthermore, Belykh et al. investigated the existence and the stability of unconditional clusters, which rise does not depend on the origin of the other clusters [123]. Ma et al. proposed a coupling scheme with cooperative and competitive weight couplings to stabilize arbitrary selected cluster synchronization states in connected chaotic networks [121].

3.3. ADD SIMPLE CONTROLLERS

The methods mentioned in Section 3.2 are obtained by adjusting the coupling matrix between nodes. However, when the network size is very big, there are more cluster synchronization manifolds, how to select the appropriate coupling matrix for one cluster synchronization manifold becomes very difficult. Therefore, we proposed a novel method to make the network achieve cluster synchronization by adding simple controllers [119].

The evolution of the dynamical variable is written as follows:

$$\dot{x}_i(t) = f(x_i(t),t) + c\sum_{j=1}^{N} a_{ij}x_j(t) + u_i(t), \quad i=1,2,\cdots,N, \qquad (3\text{-}4)$$

where $u_i(t)$ is a control input added on node i. The definitions of the rest parameters are the same as those in (1-6).

The coupling matrix $A = (a_{ij})_{N\times N} \in R^{N\times N}$ represents the coupling configuration of the network. Suppose the network is connected in the sense that there are no isolated clusters in the network, the diffusively coupled matrix A is symmetric and irreducible with its eigenvalues being ordered as $0 = \lambda_1(A) > \lambda_2(A) \geq \cdots \geq \lambda_N(A)$.

In the following, we investigate a new approach to make a complex dynamical network (3-4) to achieve cluster synchronization.

Assume a network is made up of $M(M \geq 2)$ groups, and the group network sizes are $N_1, N_2, \cdots$, and N_M, respectively, with $N_1 + N_2 + \cdots + N_M = N$. Label the nodes from number 1 to number N without change of network topology. Without loss of generality, we assume that group 1 includes the nodes from number 1 to number N_1; group 2 includes the nodes from number $N_1 + 1$ to number $N_1 + N_2$, and so on.

The cluster synchronization manifold is defined as:

$$\begin{cases} x_1(t) = x_2(t) = \cdots = x_{N_1}(t) \\ x_{N_1+1}(t) = x_{N_1+2}(t) = \cdots = x_{N_1+N_2}(t) \\ \vdots \\ x_{N_1+\cdots+N_{M-1}+1}(t) = x_{N_1+\cdots+N_{M-1}+2}(t) = \cdots = x_{N_1+\cdots+N_{M-1}+N_M}(t) \end{cases} \qquad (3\text{-}5)$$

The network locally/globally reaches clustering synchronization manifold if the nodes in the same group locally/ globally reaches the synchronization manifold corresponding to the corresponding group.

In order to make the network achieve cluster synchronization, we adopt a new approach as follows:

$$u_i(t) = -c \sum_{j=1, j \neq i}^{N} a_{ij}(s_i(t) - s_j(t)) = -c \sum_{j=1}^{N} a_{ij} s_j(t) \quad (3\text{-}6)$$

Here $s_i(t)$ is the desired state of node i at time t. Obviously, if node i and node j are included in a same group, then $s_i(t) = s_j(t)$ and the contribution of the link between them to the controller $u_i(t)$ is zero. Furthermore, when there is only one group in the network, the controller $u_i(t)$ is zero.

In this paper, the desired state $s_i(t)$ of node i at time t is chosen as the average states of all the nodes in the same group at time t as followings:

$$\begin{cases} s_1(t) = s_2(t) = \cdots = s_{N_1}(t) = \dfrac{x_1(t) + \cdots + x_{N_1}(t)}{N_1} \\ s_{N_1+1}(t) = s_{N_1+2}(t) = \cdots = s_{N_1+N_2}(t) = \dfrac{x_{N_1+1}(t) + \cdots + x_{N_1+N_2}(t)}{N_2} \\ \vdots \\ s_{N_1+\cdots+N_{p-1}+1}(t) = s_{N_1+\cdots+N_{M-1}+2}(t) = \cdots = s_N(t) = = \dfrac{x_{N_1+\cdots+N_{M-1}+1}(t) + \cdots + x_N(t)}{N_M} \end{cases} \quad (3\text{-}7)$$

and

$$\dot{s}_i(t) = f(s_i(t), t) \qquad i = 1, 2, \cdots, N \quad (3\text{-}8)$$

The local stability and global stability with the desired states satisfying the equation (3-7) and equation (3-8) are introduced as following.

3.3.1. Local Stability Analysis

When the controller u_i is selected as the equation (3-6) and the synchronized state of each group is selected as the equation (3-7), the local stability of cluster synchronization is investigated.

Theorem 3-1[119]: Assume f is continuously differentiable and A is a diffusively coupled matrix, the network (3-4) locally reaches cluster synchronization if and only if the linear time-varying systems

$$\frac{d\delta y_k(t)}{dt} = \delta y_k(t)\left(Df(s_k(t),t) + c\lambda_k(A)I\right) \qquad k = 2,\cdots,N \tag{3-9}$$

are exponentially stable about the zero solution.

Proof: Denote $\delta x_i(t) = x_i(t) - s_i(t)$, according to the equation (3-7) and (3-8), differentiate the isolated dynamics of the node i along $s_i(t)$, then

$$\begin{aligned}\frac{d\delta x_i(t)}{dt} &= f(x_i(t),t) - f(s_i(t),t) + c\sum_{j=1}^{N} a_{ij} x_j(t) - c\sum_{j=1}^{N} a_{ij} s_j(t) \\ &= Df(s_i(t),t)\delta x_i(t) + c\sum_{j=1}^{N} a_{ij}\delta x_j(t)\end{aligned} \tag{3-10}$$

where $Df(s_i(t),t)$ is the Jacobin matrix along $s_i(t)$ of $f(x,t)$.

Let $Df(s(t),t) = Diag[Df(s_1(t),t),\cdots,Df(s_N(t),t)]$ and $\delta x(t) = [\delta x_1(t),\cdots,\delta x_N(t)]$, then

$$\frac{d\delta x(t)}{dt} = \delta x(t) Df(s(t),t) + c\delta x(t) A \tag{3-11}$$

Since A is a real symmetric and irreducible matrix, then the matrix A can be decomposed as: $A = WJW^T$, where $J = diag[\lambda_1(A), \cdots, \lambda_N(A)]$, and $0 = \lambda_1(A) > \lambda_2(A) \geq \cdots \geq \lambda_N(A)$.

Let $\delta y(t) = \delta x(t)W$, then we have

$$\frac{d\delta y_k(t)}{dt} = \delta y_k(t)\left(Df(s_k(t),t) + c\lambda_k(A)I\right) \qquad k = 1, 2, \cdots, N \tag{3-12}$$

Therefore, the stability problem of cluster synchronization state is transferred to the stability problem of the N pieces of n-dimensional linear time-varying system (3-11). Because $\lambda_1 = 0$ corresponds to cluster synchronization state of the system. Therefore, if the following $N-1$ pieces of n-dimensional linear time variable systems are exponentially stable:

$$\frac{d\delta y_k(t)}{dt} = \delta y_k(t)\left(Df(s_k(t),t) + c\lambda_k(A)I\right) \qquad k = 2, \cdots, N \tag{3-13}$$

then the cluster synchronization state is exponentially stable.
The proof is complete.

Corollary 3-1[119]: Assume f is continuously differentiable and chaotic, the network (3-4) locally reaches cluster synchronization if

$$c > h_{max} / |\lambda_2(A)| \tag{3-14}$$

where $h_i (i = 1, \cdots, n)$ denote the corresponding Lyapunov exponents of each isolated n-dimensional dynamical node and $h_{max} = \max\limits_{i=1,\cdots,n} h_i$ is positive.

Proof: As known from theorem 1, the network locally reaches cluster synchronization if the equation (3-9) satisfies.

Since f is chaotic, which means that for each $\lambda_k \neq 0$, if the transversal Lyapunov exponents are all negative, then the cluster synchronization state is exponentially stable. Since A is a diffusively coupled matrix, then its eigenvalues can be ordered as $0 = \lambda_1(A) > \lambda_2(A) \geq \cdots \geq \lambda_N(A)$. It implies that

$$h_{max} + c\lambda_2(A) < 0 \qquad (3\text{-}15)$$

That is, the stability condition of the cluster synchronization state is

$$c > h_{max} / |\lambda_2(A)|$$

The proof is complete.

3.3.2. Global Stability Analysis

Theorem 3-2[119]. Assume f is continuously differentiable and A is a diffusively coupled matrix. If there exist positive definite matrices $P = diag[p_1, \cdots, p_n]$, $\Delta = diag[\Delta_1, \cdots, \Delta_n]$ and a constant $\eta > 0$, such that $(x-y)^T P(f(x,t) - f(y,t) + \Delta y - \Delta x) \leq -\eta(x-y)^T(x-y)$ for all $x, y \in R^n$ and $\Delta_{max} + c\lambda_2(A) < 0$, where $\Delta_{max} = \max\limits_{k=1,\cdots,n} \Delta_k$, then the network (3-4) with controller (3-6) is globally exponentially stabilized to cluster synchronization state.

Proof: Define a Lyapunov function as

$$V(t) = \frac{1}{2}\sum_{i=1}^{N} \delta x_i(t)^T P \delta x_i(t) \qquad (3\text{-}16)$$

Denote $\tilde{x}^k(t) = [\delta x_{1k}(t), \cdots, \delta x_{Nk}(t)]^T$, $k = 1, \cdots n$. Then, we have

$$\frac{dV(t)}{dt} = \sum_{i=1}^{N} \delta x_i(t)^T P \frac{d\delta x_i(t)}{dt}$$

$$= \sum_{i=1}^{N} \delta x_i(t)^T P[f(x_i(t),t) - f(s_i(t),t) + c\sum_{j=1}^{N} a_{ij}\delta x_j(t)]$$

$$= \sum_{i=1}^{N} \delta x_i(t)^T P[f(x_i(t),t) - f(s_i(t),t) - \Delta\delta x_i(t)]$$

$$+ \sum_{i=1}^{N} \delta x_i(t)^T P[c\sum_{j=1}^{N} a_{ij}\delta x_j(t) + \Delta\delta x_i(t)]$$

$$\leq -\eta \sum_{i=1}^{N} \delta x_i(t)^T \delta x_i(t) + \sum_{i=1}^{N} \delta x_i(t)^T P[c\sum_{j=1}^{N} a_{ij}\delta x_j(t) + \Delta\delta x_i(t)]$$

$$= -\eta \sum_{i=1}^{N} \delta x_i(t)^T \delta x_i(t) + \sum_{k=1}^{n} p_k \tilde{x}^k(t)^T (cA + \Delta_k I)\tilde{x}^k(t)$$

Because $\Delta_{max} + c\lambda_2(A) < 0$ and $0 > \lambda_1(A) \geq \cdots \geq \lambda_N(A)$, then the matrix $cA + \Delta_k I$ is negative definite. So we have

$$\frac{dV(t)}{dt} \leq -\eta \sum_{i=1}^{N} \delta x_i(t)^T \delta x_i(t) \leq -\eta \frac{V(t)}{\min_i p_i}$$

Therefore

$$V(t) = o(e^{\frac{-\eta t}{\min_i p_i}}).$$

The proof is complete.

3.3.3. Simulation Results

As a typical example, the Chua's circuit is considered as each isolated node's dynamics in the complex dynamical network:

$$\begin{cases} \dot{y}_1 = k[y_2 - h(y_1)] \\ \dot{y}_2 = y_1 - y_2 + y_3 \\ \dot{y}_3 = -ly_2 \end{cases} \tag{3-17}$$

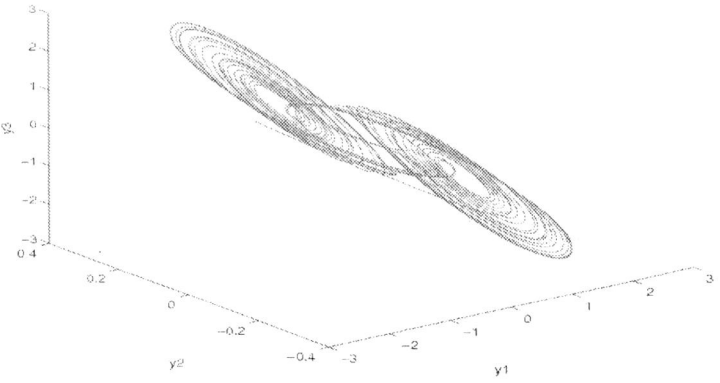

Figure 3.2. Chaotic behavior of Chua's circuit(figure taken from [119], © Communications in Theoretical Physics).

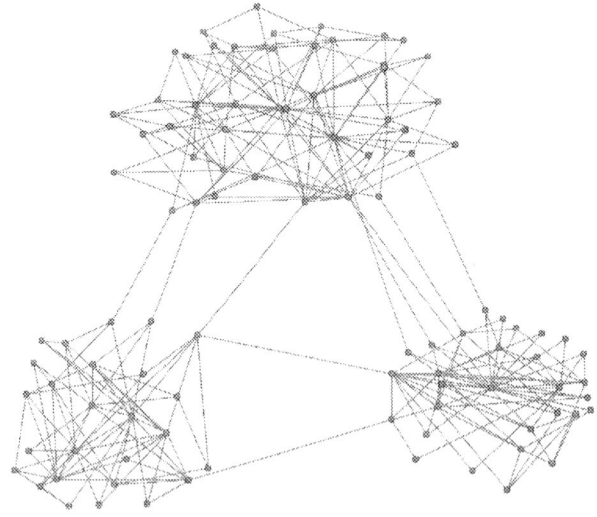

Figure 3.3. A network with size $N = 100$ and three groups (figure taken from [119] ©Communications in Theoretical Physics).

where $h(x) = \frac{2}{7}x - \frac{3}{14}[|x+1| - |x-1|]$, $k = 9$, and $l = 14\frac{2}{7}$. Under these parameters, the Chua's circuit has a double-scroll chaotic attractor, as shown in Figure 3.2.

In this simulation, the algorithm in [124] is used to generate a network with three groups as shown in Figure 3.3. These three group sizes are 28, 34 and 38, respectively and the weight of each edge is assigned to 1. The nodes' dynamic is a Chua circuit oscillator and the initial states of the nodes are randomly distributed in [-0.3 -0.1]. It is known that when $P = I_3$, $\Delta = 10I_3$, $\eta = 0.6218$ [98], that is

$$(x-y)^T P(f(x,t) - f(y,t) + \Delta y - \Delta x) \leq -\eta(x-y)^T(x-y) \qquad (3\text{-}18)$$

A controller is added to each node as the equation (3-6). Here the coupling strength $c = 50$ and $\lambda_2(A) = -0.2642$, then in-equation (3-18) and $\Delta_{\max} + c\lambda_2(A) < 0$ satisfies. Therefore, according to theorem 2, the network can globally reach cluster synchronization. As can be seen from Figure 3, for $t > 5s$, 28 red dotted lines reduces to one red dotted line, 34 magenta dotted lines reduces to one magenta dotted line, and the rest 38 blue solid lines reduces to one blue solid line. This means that the network reach the desired cluster synchronization. Furthermore, these three curves are chaotic.

From the controller (3-6) and the simulation result in the Figure 3.4, we can know that when each dynamic node only know its neighbor s' destinations and its destinations, the whole network can achieve different destinations. In other words, the information which each node need to know is very small, which makes the approach can be used to applications.

This provides us a novel approach to how to make a network completes different goals. For example, we can use this means to make the network laser jets to form different beams of laser light. When each node represents a moving robot, the novel approach can make these robots achieve different goals such as saving few people from the fire and other dangerous places.

3.4. ADAPTIVE CLUSTER SYNCHRONIZATION OF COMPLEX NETWORKS

As we know from Section 3.2 and 3.3, the network may reach cluster synchronization only if the coupling strength between connected nodes is bigger than some critical coupling value. However, how to select the coupling strength value is difficult, especially when the network size is bigger. Combined with the contents in Chapter 2, an adaptive approach based on the local information is proposed to obtain the critical coupling value [120].

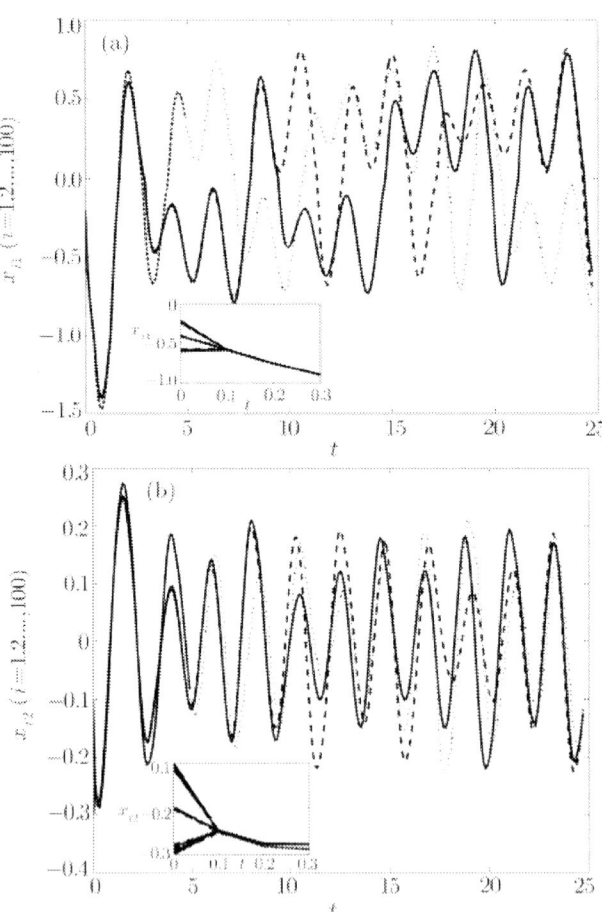

Figure 3.4 (Continued)

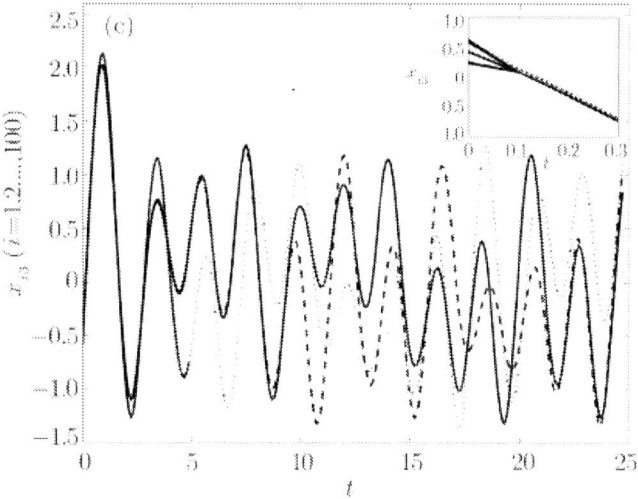

Figure 3.4. Cluster synchronization in the network shown in Figure 3.2 with the node's dynamic being a chaotic Chua circuit oscillator and coupling strength $c = 50$: $x_{i1}(t)$ (top), $x_{i2}(t)$ (middle) and $x_{i3}(t)$ (bottom) (figure taken from [119], ©Communications in Theoretical Physics).

Consider a complex dynamical network of N linearly diffusively coupled identical nodes, and the evolution of the dynamical variable is written as follows:

$$\dot{x}_i(t) = f(x_i(t), t) + \sum_{j=1}^{N} a_{ij}(t) x_j(t) + u_i(t), \quad i = 1, 2, \cdots, N, \quad (3\text{-}19)$$

The definitions of the parameters are the same as those in (3-4).

3.4.1. Adaptive Strategy in Cluster Synchronization

In order to make the network achieve cluster synchronization (3-5), the input is designed as follows[120]:

$$u_i(t) = -\sum_{j=1, j\neq i}^{N} a_{ij}(t)(s_i(t) - s_j(t)) \quad (3\text{-}20)$$

where the coupling strength $a_{ij}(t) = (a_{ij}^1(t), \cdots, a_{ij}^n(t))^T$ between nodes adopts the following adaptive strategy.

$$\dot{a}_{ij}(t) = \beta \|(x_j(t) - s_j(t)) - (x_i(t) - s_i(t))\| \quad (3\text{-}21)$$

where $\beta > 0$ is the adaptive gain. Obviously, when the adaptive gain β is zero and there is only one group in the network, the destination state is equal for any random node i and node j, which means that $s_i(t) = s_j(t)$. Therefore, the equation (3-21) reduces to

$$\dot{a}_{ij}(t) = \beta \|x_j(t) - x_i(t)\|$$

which is the edge-strategy in [93].

3.4.2. Global Stability Analysis of Cluster Synchronization

When the controller u_i is designed as the equation (3-20) and the adaptive strategy is chosen as the equation (3-21), the global stability of cluster synchronization state (3-5) is investigated.

Theorem 3-3[120]. Suppose that there exist positive definite matrices $P = diag[p_1, \cdots, p_n]$, $\Delta = diag[\Delta_1, \cdots, \Delta_n]$ and a constant $\eta > 0$, such that

$$(x-y)^T P(f(x,t) - f(y,t) + \Delta y - \Delta x) \leq -\eta(x-y)^T(x-y) \quad \forall x, y \in R^n, \quad (3\text{-}22)$$

and $\Delta_{max} + \lambda_2(A) < 0$, where $\Delta_{max} = \max_{k=1,\cdots,n} \Delta_k$. Then the network globally reaches cluster synchronization state (3-5) with the controller (3-20) and (3-21).

Proof: Define a Lyapunov function as

$$V(t) = \frac{1}{2}\sum_{i=1}^{N}\delta x_i(t)^T P \delta x_i(t) + \frac{1}{2\beta}\sum_{\varepsilon}(c_{ij}-a_{ij})^T(c_{ij}-a_{ij}). \tag{3-23}$$

where $\delta x_i(t) = x_i(t) - s_i(t)$, $c_{ij} = [c_{ij}^1, c_{ij}^2, \cdots, c_{ij}^n]$ is also n-dimension.

Denote $\tilde{x}^k(t) = [\delta x_{1k}(t), \cdots, \delta x_{Nk}(t)]^T$, $k=1,\cdots n$. Then, it is known that

$$\frac{dV(t)}{dt} = \sum_{i=1}^{N}\delta x_i(t)^T P \frac{d\delta x_i(t)}{dt} - \frac{1}{\beta}\sum_{\varepsilon}(c_{ij}-a_{ij})^T \dot{a}_{ij}$$

$$= \sum_{i=1}^{N}\delta x_i(t)^T P[f(x_i(t),t) - f(s_i) + \sum_{j=1}^{N}a_{ij}\delta x_j(t) - \frac{1}{\beta}\sum_{\varepsilon}(c_{ij}-a_{ij})^T \dot{a}_{ij}$$

$$= \sum_{i=1}^{N}\delta x_i(t)^T P[f(x_i(t),t) - f(s_i) + \sum_{j=1}^{N}a_{ij}\delta x_j(t) - \sum_{\varepsilon}(c_{ij}-a_{ij})^T \|x_j - s_j - x_i + s_i\|$$

$$= \sum_{i=1}^{N}\delta x_i(t)^T P[f(x_i(t),t) - f(s_i) - \Delta \delta x_i(t)]$$

$$+ \sum_{i=1}^{N}\delta x_i(t)^T P[\sum_{j=1}^{N}a_{ij}\delta x_j(t) + \Delta \delta x_i(t)] - \sum_{\varepsilon}(c_{ij}-a_{ij})^T \|x_j - s_j - x_i + s_i\|$$

$$\leq -\eta \sum_{i=1}^{N}\delta x_i(t)^T \delta x_i(t) + \sum_{i=1}^{N}\delta x_i(t)^T P[\sum_{j=1}^{N}a_{ij}\delta x_j(t) + \Delta \delta x_i(t)]$$

$$- \sum_{\varepsilon}(c_{ij}-a_{ij})^T \|x_j - s_j - x_i + s_i\|$$

$$= -\eta \sum_{i=1}^{N}\delta x_i(t)^T \delta x_i(t) + \sum_{k=1}^{n}p_k \tilde{x}^k(t)^T (A+\Delta_k I) \tilde{x}^k(t) - \sum_{\varepsilon}(c_{ij}-a_{ij})^T \|x_j - s_j - x_i + s_i\|$$

Since $\Delta_{max} + \lambda_2(A) < 0$ and $0 > \lambda_2(A) \geq \cdots \geq \lambda_N(A)$, the matrix $A + \Delta_k I$ is negative semi-definite. Furthermore, for any edge $(i,j) \in \varepsilon$, there exists the value c_{ij} is bigger than the corresponding edge strength a_{ij}, it is obtained that

$$\frac{dV(t)}{dt} \leq -\eta \sum_{i=1}^{N}\delta x_i(t)^T \delta x_i(t) \leq -\eta \frac{V(t)}{\max_i p_i}. \tag{3-24}$$

Therefore,

$$V(t) \leq \exp\left\{\frac{-\eta t}{\max_i p_i}\right\} V(0). \tag{3-25}$$

The proof is complete.

3.4.3. Simulation Results

A selected network is divided into three groups, which means that $M = 3$. The desired state $s_i(t)$ of node i at time t is same as (3-7). As a known chaotic oscillator, the Rossler oscillator is considered as each isolated node's dynamics in the complex dynamical network:

$$\dot{y}_1 = -y_2 - y_3 \tag{3-26}$$

$$\dot{y}_2 = y_1 + 0.2 y_2 \tag{3-27}$$

$$\dot{y}_3 = 0.2 + z(x - 5.7) \tag{3-28}$$

The nodes initial states are randomly distributed in [-5 5]. From the reference [5], it is known that the Rossler oscillator satisfies the in-equation (3-22) in the Theorem 3-3.

3.4.3.1. BA Scale-free Network without Noise

At first, a BA scale-free network is constructed with $m = m_0 = 5$ and $N = 80$. Here three group sizes N_1, N_2 and N_3 are 24, 38 and 18, respectively. The distributed controller (3-20) is added to each node with the edge strength is adaptive according to (3-21). For $t > 11s$, the in-equation $\Delta_{\max} + \lambda_2(A) < 0$ can be easily verified by calculation, then the network can be globally stabilized at the cluster synchronization state in theory. As can be seen from Figure 3.5, for $t > 11s$, 24 dashed lines reduces to one dashed line, 38 dotted lines reduces to one dotted line, and the rest 18 solid lines reduces to one solid line. This means that the network reaches the desired cluster synchronization on each node's three dimensions. In Figure 3.6, for the edge strength a_{ij} between node i and node j, its all dimensions keep constant for $t > 11s$. Combined with Figure 3.5, it is found that when the edge strengths are fixed, the network reaches cluster synchronization.

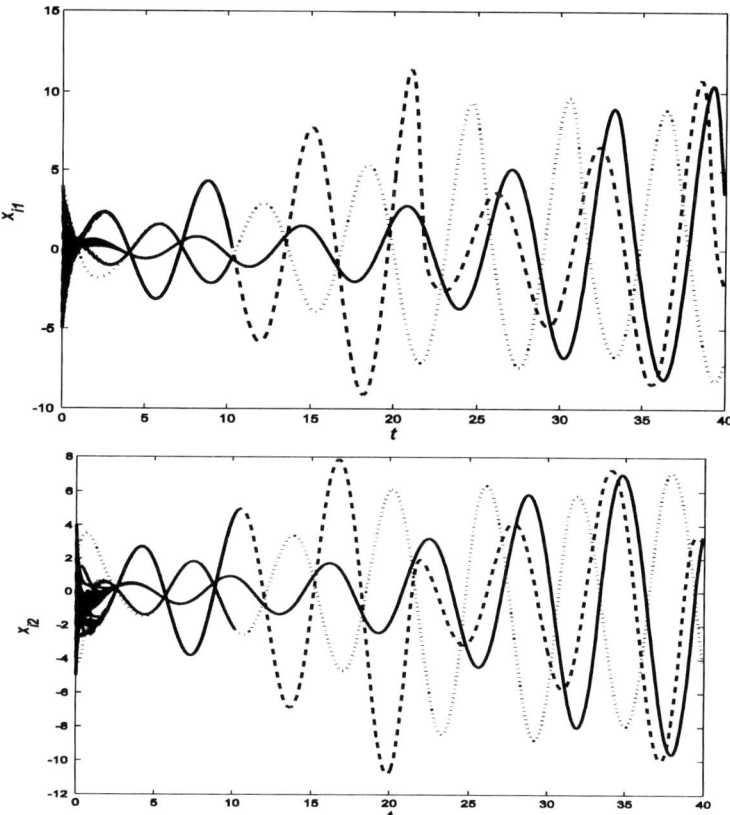

Figure 3.5. (Continued)

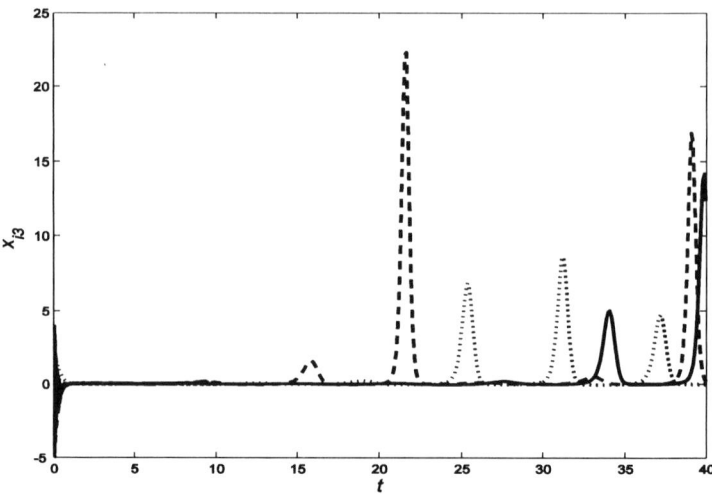

Figure 3.5. Cluster synchronization in the BA scale-free network with the node being a chaotic Rossler oscillator and the adaptive gain $\beta = 0.1$: $x_{i1}(t)$ (top), $x_{i2}(t)$ (middle) and $x_{i3}(t)$ (bottom) (figure taken from [120] ©Elsevier).

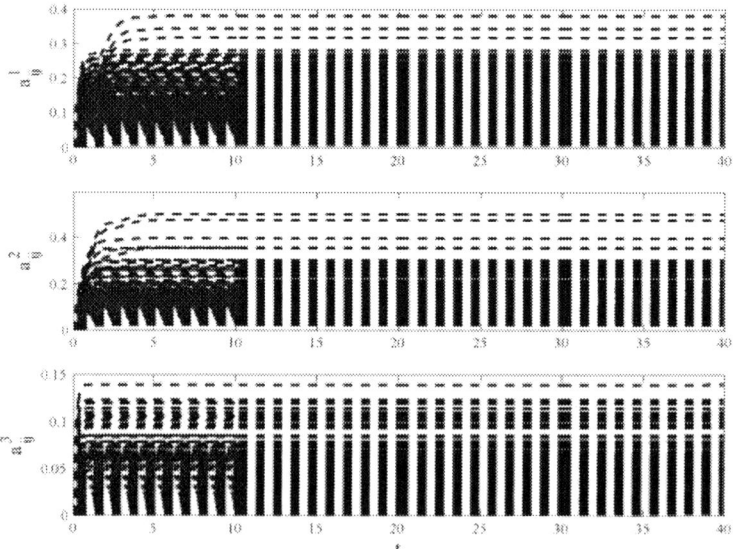

Figure 3.6. Edge strengths' evolution during the interval of the BA scale-free network in Figure 3.5 reaches cluster synchronization (figure taken from [120] ©Elsevier).

3.4.3.2. BA Scale-free Network with Noise

In the above simulations, noise, which often exists in the real world networks, is not considered. An independent noise $D\psi_i$ is added to the variable of the equation (3-26)-(3-28) respectively at $t = 40s$ in Figure 3.5, where the amplitude of the noise D is 0.5, 0.3 and 0.1; ψ_i randomly satisfies the normal Gaussian distribution $N(0,1)$. As can be seen from Figure 3.7 and Figure 3.8, the network reaches cluster synchronization and the coupling strength between nodes is fixed for $t > 43s$. This means the adaptive strategy can reduce the impact of noise.

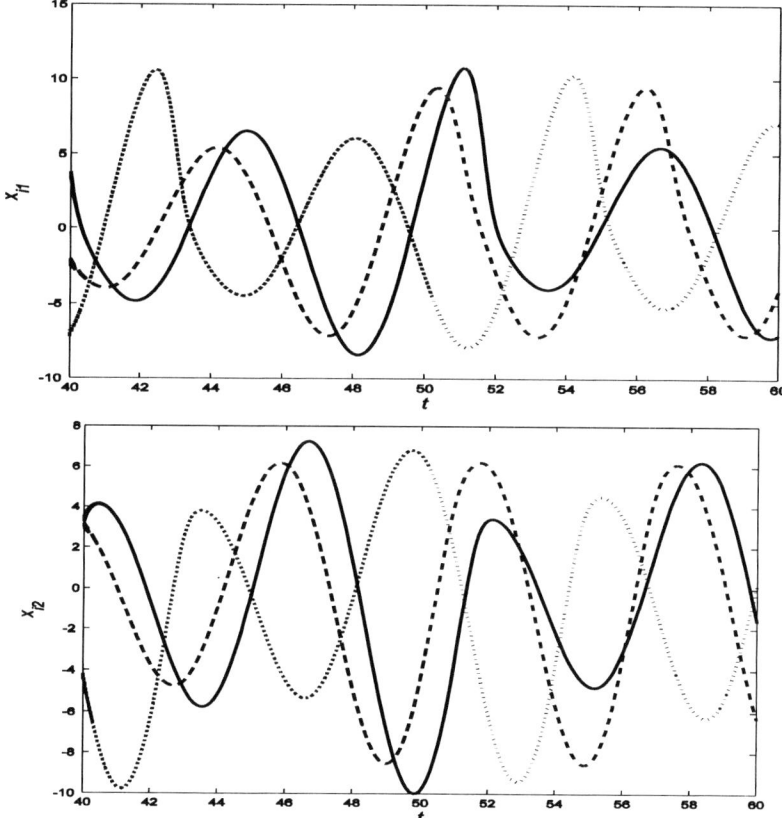

Figure 3.7. (Continued)

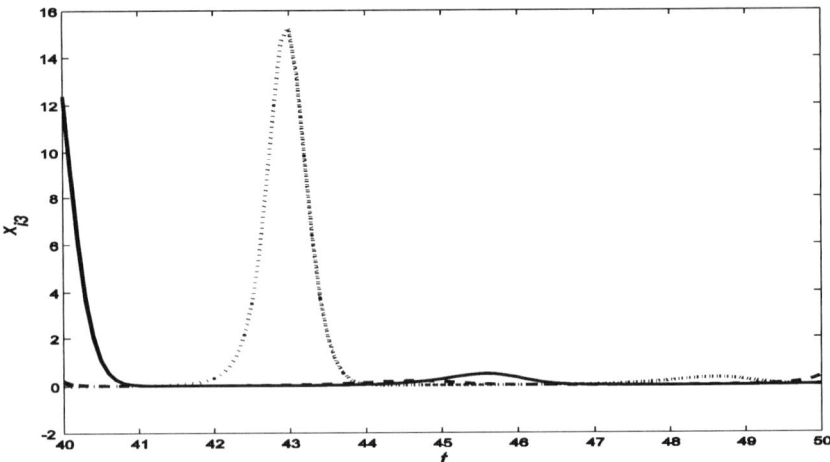

Figure 3.7. Cluster synchronization in the BA scale-free network in Figure 3.5 when the noise is added at $t = 40s$: $x_{i1}(t)$ (top), $x_{i2}(t)$ (middle) and $x_{i3}(t)$ (bottom) (figure taken from [120] ©Elsevier).

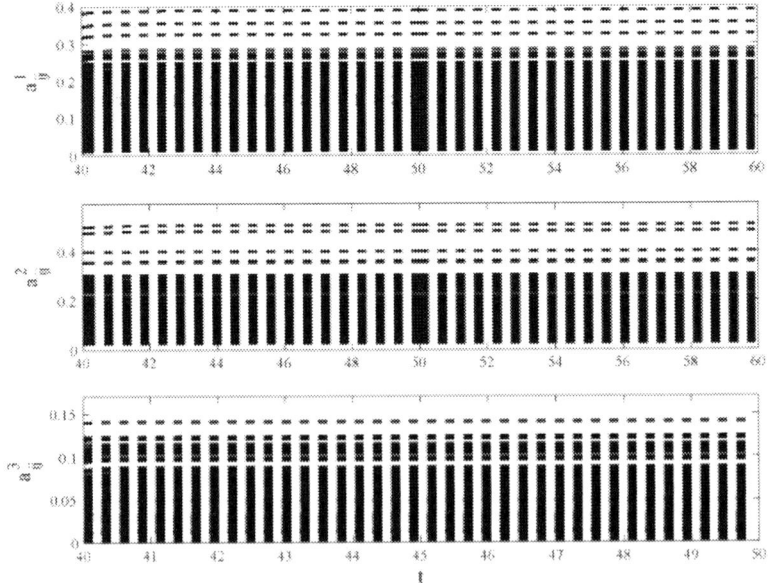

Figure 3.8. Edge strengths' evolution during the interval of the BA scale-free m network in Figure 3.7 reaches cluster synchronization (figure taken from [120] ©Elsevier).

3.4.3. Nonidentical Oscillators

Now, the cluster synchronization of network with slightly oscillators is considered. A parameter η_i, which is randomly distributed in [0.9, 1.1], is added to (3-19) as follows [125]:

$$\dot{x}_i(t) = \eta_i f(x_i(t), t) + \sum_{j=1, j \neq i}^{N} a_{ij}(t)(x_j(t) - x_i(t)) + u_i(t), \quad i = 1, 2, \cdots, N,$$

(3-29)

In this section, a BA scale-free network is used with network size $N = 100$ and three group sizes N_1, N_2 and N_3 being 24, 58 and 18, respectively. Figure 3-9 shows that for $t > 5s$, 24 dashed lines almost reduces to one dashed line, 58 dotted lines almost reduces to one dotted line, and the rest 18 solid lines almost reduces to one solid line. However, the network can not reach cluster synchronization until $t = 40s$. Combined with the change of coupling strength between nodes in Figure 3.10, it is easy to find the coupling strength increases slowly from $t = 5s$ to $t = 40s$. This means that left side of (3-20) is not equal to zero. Therefore, the desired states of nodes cannot achieve.

As we know, the BA scale-free network has heterogeneous degree distribution. In the Figure 3.11, the impacts of degree are investigated at different time. At time t, the synchronization difference of node i is defined as $\delta x_i(t) = \|x_i(t) - s(t)\|$; an average difference $\delta x(k)$ is average of the synchronization difference of nodes with degree being k. For $t = 2s$, it is can be found form Figure 3.10 that the network has weak coupling strength between nodes. However, it is shown in Figure 3.11, the node with degree 3 has large average difference; while the average difference of the nodes with large degree is very low. This implies that the network shows the hierarchical synchronization. As time increases, Figure 3.10 shows that the coupling strength increases, this phenomenon no longer exists. For $t = 40s$, the average differences for all nodes are nearly to be zero, which means that the network nearly reaches cluster synchronization.

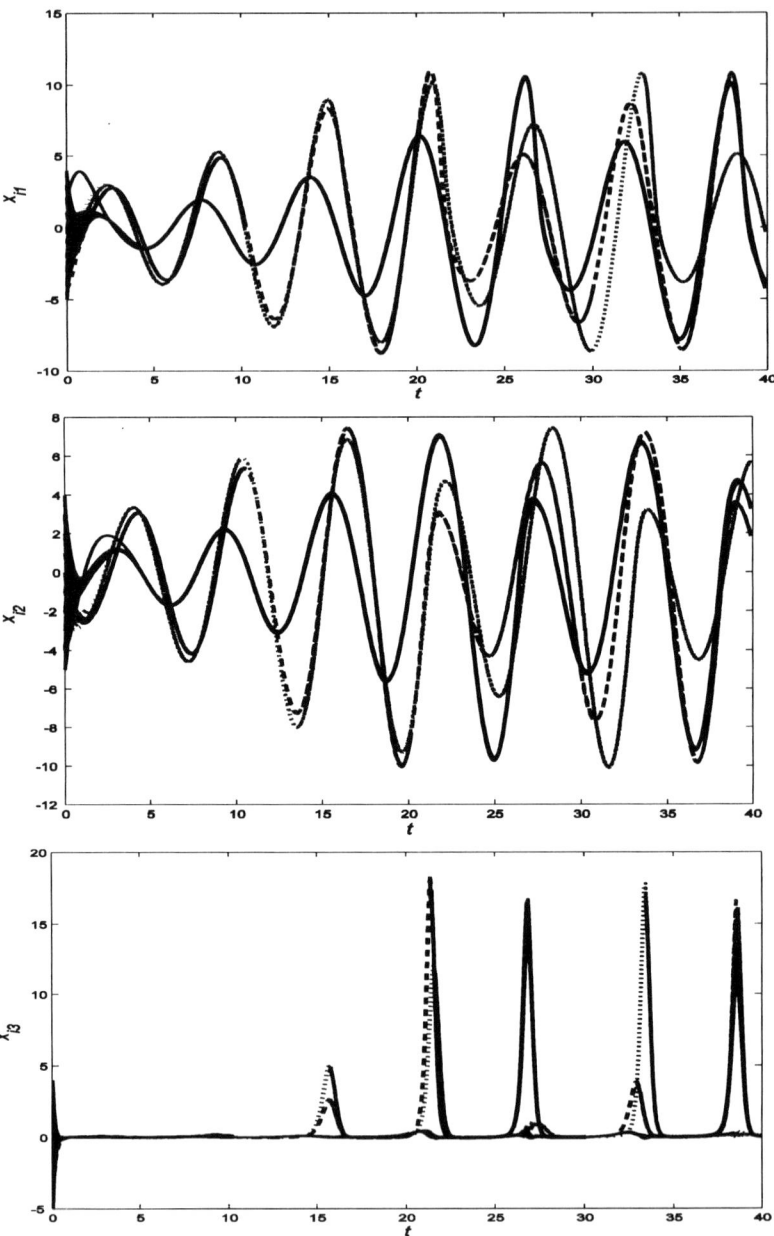

Figure 3.9. Cluster synchronization in the BA scale-free network with network size $N = 100$: $x_{i1}(t)$ (top), $x_{i2}(t)$ (middle) and $x_{i3}(t)$ (bottom) (figure taken from [120] ©Elsevier).

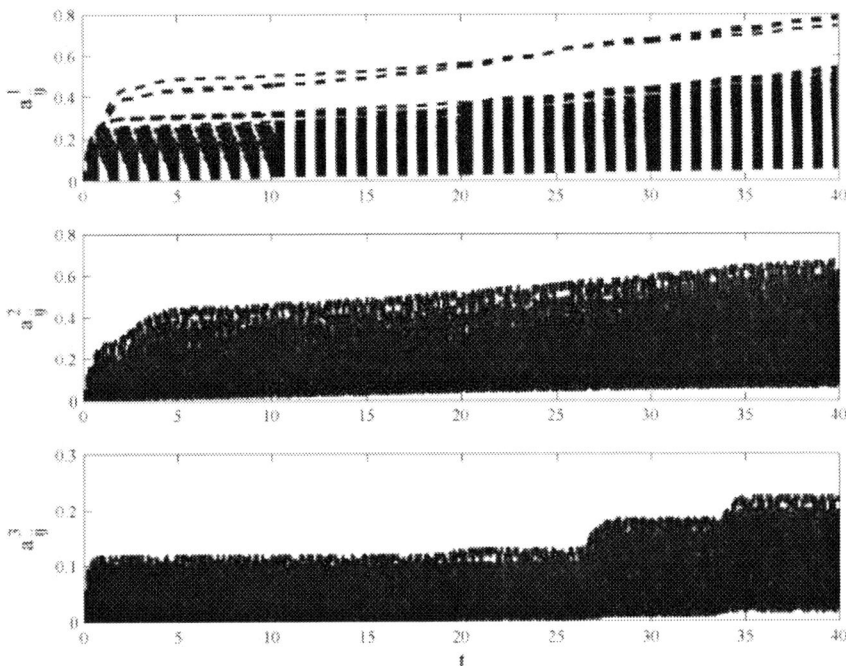

Figure 3.10. Edge strengths' evolution during the interval of the BA scale-free network in Figure 3.9 reaches cluster synchronization (figure taken from [120] ©Elsevier).

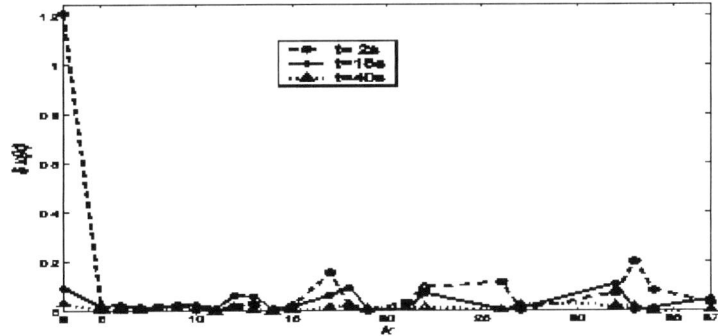

Figure 3.11. Cluster synchronization difference $\delta x(k)$ of the BA scale-free network in Figure 3.9 at different time (figure taken from [120] ©Elsevier).

Chapter 4

CONTROL OF COMPLEX DYNAMICAL NETWORKS

ABSTRACT

In this chapter, the control of equilibrium point in complex network is presented. At first, a pinning control method is proposed to make a BA scale-free network achieve an equilibrium point of an isolated node. The pinning control means that by adding feedback controllers to few nodes in the network, the whole nodes of the network may achieve the desired state. For a strong connected network, it is found that by adding controller to only one node, the network achieves the desired equilibrium point. At last, the method to make the network achieve a heterogeneous equilibrium point is given, in which the whole nodes are divided into several subgroups, the nodes in the same subgroup achieve one equilibrium point of the isolated node, while the nodes in different subgroups achieve different equilibrium points.

4.1. INTRODUCTION

As the continuous increase of network size and the complexity of nonlinear dynamical characteristics, how to control of these dynamical networks becomes an important topic. Pinning control strategy is proofed to be an effective approach. Pinning control was first proposed to investigate the chaotic systems, which is made up of mutually coupled chaotic nodes[126,

127]. The principle of the pinning control approach is that when few nodes of the whole network are added controllers, all the nodes are controlled.

4.2. CONTROL A GENERAL DYNAMICAL NETWORK TO A HOMOGENEOUS EQUILIBRIUM POINT

Wang and Chen first applied the pinning control strategy to control of the scale-free network[128]. The main result is that by adding few controllers to few nodes of scale-free network, all the nodes may be controlled to an equilibrium point of the isolated node, which means that the network may reach a homogeneous equilibrium point. Furthermore, Li et al. extended the above pinning control method to networks with more general topologies[129].

Consider a dynamical network of N linearly and diffusively coupled identical oscillators, with each oscillator being an n-dimensional dynamical system. The evolution of the dynamical variables is written in the form

$$\dot{x}_i = f(x_i) + \sum_{j=1, j \neq i}^{N} c_{ij} a_{ij} H(x_j - x_i), \quad i = 1, 2, \cdots, N \quad (4\text{-}1)$$

where $c_{ij} > 0$ represents the coupling strength between node i and node j. The definitions of the rest parameters are the same as those in (1-6).

Suppose the object is to control the network to an equilibrium point $\bar{x}$, which means that $f(\bar{x}) = 0$. The designed pinning control strategy is as follows:

$$\begin{cases} \dot{x}_i = f(x_i) + \sum_{j=1, j \neq i}^{N} c_{ij} a_{ij} H(x_j - x_i) - c_{ii} d_i H(x_i - \bar{x}), & i = 1, 2, \cdots, l \\ \dot{x}_i = f(x_i) + \sum_{j=1, j \neq i}^{N} c_{ij} a_{ij} H(x_j - x_i), & i = l+1, l+2, \cdots, N \end{cases}$$

$$(4\text{-}2)$$

where the control gain $d_i > 0$ and the coupling strength between nodes satisfies

$$c_{ii}a_{ii} + \sum_{j=1, j \neq i}^{N} c_{ij}a_{ij} = 0$$

Since the number of the nodes may be renumbered, it is easy to define the former l nodes are added feedback controllers. Suppose two diagonal matrices

$$D = diag\{d_1, d_2, \ldots d_l, 0, \ldots 0\} \in R^{N \times N}$$

$$D' = diag\{c_{11}d_1, c_{22}d_2, \ldots c_{ll}d_l, 0, \ldots, 0\} \in R^{N \times N}.$$

Then the network (4-2) may be rewritten as follows:

$$\begin{aligned}\dot{x} &= f(x) - \left[(G+D) \otimes H\right]x + (D' \otimes H)\bar{X} \\ &= I_N \otimes f(x_i) - \left[(G+D) \otimes H\right]x + (D' \otimes H)\bar{X}\end{aligned} \quad (4\text{-}3)$$

where $\otimes$ represents the Kronecker product, the matrix the entry of G is $g_{ij} = -c_{ij}a_{ij}$. Then the matrix G is semi-definite, and the matrix $G+D$ is definite with its minimum eigenvalue $\lambda_{\min}(G+D) > 0$.

Definition 4-1[130] A function $f(x,t)$ is Lipschitz continuously and its Lipschitz constant is L_c^f, if

$$\|f(x,t) - f(y,t)\| \leq L_c^f \|x - y\| \quad (4\text{-}4)$$

for all x, y, t.

Definition 4-2[130] If there exists $\delta > 0$, the inequation $\|x_i(0) - x_j(0)\| \leq \delta$ satisfies for any random $i, j = 1, \cdots, N$. Moreover, there exists $\varepsilon > 0$, $T > 0$ and $K > 0$ for all $t > T$ and $i,j = 1,\cdots,N$, the inequation

$$\|x_i(t) - x_j(t)\| \le Ke^{-\varepsilon t} \tag{4-5}$$

satisfies, then the network (4-1) is said to achieve local synchronization.

Definition 4-3[130] If there exists $\varepsilon > 0$, $T > 0$ and $K > 0$ for all $t > T$ and $i,j = 1,\cdots,N$, the inequation (4-5) satisfies, then the network (4-1) is said to achieve global synchronization.

Theorem 4-1[129]. Suppose $f(x)$ is Lipschitz continuous and the Lipschitz constant $L_c^f > 0$, and the matrix H is symmetric and positive definite. If

$$\lambda_{\min}(G+D) > \alpha \equiv L_c^f / \lambda_{\min}(H) \tag{4-6}$$

where $\lambda_{\min}(H)$ and $\lambda_{\min}(G+D)$ are the minimum eigenvalue of the matrix H and $G+D$, respectively; then the complex dynamical network (4-1) may globally stabilize on the homogeneous equilibrium point $\overline{X}$.

According to the maximum Lyapunov exponent of the isolated chaotic node, the local stability of (4-1) may be obtained.

Theorem 4-2[129]: Let the maximum Lyapunov exponent of the isolated chaotic node $\dot{x}_i = f(x_i)$ be $h_{\max} > 0$. If $c_{ij} = c$, $d_i = cd$, $H = I_n$, and

$$c > h_{\max} / \lambda_{\min}(-A + diag\{d,\ldots,d,0,\ldots,0\}) \tag{4-7}$$

Then the chaotic network (4-1) may locally stabilize on the homogeneous equilibrium point $\overline{X}$.

Consider each isolated node is Chen's chaotic system as follows:

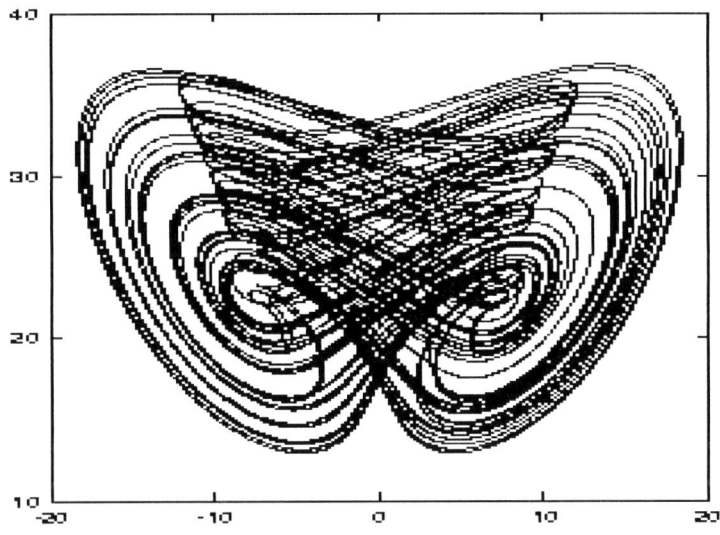

Figure 4.1. $x_1 - x_3$ of chaotic Chen's system(figure taken from [129], ©IEEE)

$$\begin{pmatrix} \dot{x}_1 \\ \dot{x}_2 \\ \dot{x}_3 \end{pmatrix} = \begin{pmatrix} 35(x_2 - x_1) \\ 25x_1 - x_1 x_3 + 28x_2 \\ x_1 x_2 - 3x_3 \end{pmatrix} \quad (4\text{-}8)$$

Then the chaotic behavior is shown in 4-1 The system (4-8) has an unstable equilibrium point $x^+ = \begin{bmatrix} 7.9373 & 7.9373 & 21 \end{bmatrix}^T$.

Suppose the object is to control all the nodes in the network to this equilibrium point x^+, the adopted pinning control strategy is

$$\begin{pmatrix} \dot{x}_{i1} \\ \dot{x}_{i2} \\ \dot{x}_{i3} \end{pmatrix} = \begin{pmatrix} p_1(x_{i2} - x_{i1}) + c \sum_{j=1}^{N} a_{ij} x_{j1} + u_{i1} \\ (p_3 - p_2) x_{i1} - x_{i1} x_{i3} + p_3 x_{i2} + c \sum_{j=1}^{N} a_{ij} x_{j2} + u_{i2} \\ x_{i1} x_{i2} - p_2 x_{i3} + c \sum_{j=1}^{N} a_{ij} x_{j3} + u_{i3} \end{pmatrix}, \quad i = 1, 2, \cdots, N$$

(4-9)

where

$$u_{ij} = \begin{cases} -cd(x_{ij} - x_j^+), & i = i_1, i_2, \cdots, i_l, j = 1, 2, 3 \\ 0 & \text{otherwise} \end{cases} \quad (4\text{-}10)$$

As we know that the scale-free network has a heterogeneous degree distribution, which means that most nodes have small degrees, while only few nodes have big degrees[15]. Therefore, there are two different pinning strategies proposed in [128]. One is the specially pinning strategy, where the nodes with biggest degrees are added controllers; the other one is random pinning strategy, where the nodes selected from the whole nodes are added controllers. Since the scale-free network has most nodes with smaller degrees, the nodes with smaller degrees have big probability to be selected.

Construct a BA scale-free network with the network size $N = 50$, and each isolated node is Chen's system. Compare the results in Figure 4.2 (a) and (b), it is easy to find that the desired coupling strength c of specially pinning two nodes with biggest degrees is smaller than those of specially pinning one node with biggest degree. Moreover, as can be seen from Figure 4.2, the time of the nodes with smallest degree reaching the desired equilibrium point is bigger than that of the nodes with biggest degree. In other words, when the node with smallest degree reaches the desired equilibrium point, the whole nodes in the network reaches the desired equilibrium point. Figure 4.3 shows that the time of the network, adopted the specially pinning strategy, reaching the desired equilibrium point is smaller than that of the randomly pinning strategy.

Recently, the pinning control strategy is extended to network with coupling matrix, which cannot be diagonalized, and the weighted networks with time delay[131, 132].

4.3. CONTROL A GENERAL DYNAMICAL NETWORK TO SYNCHRONIZATION STATE

The pinning control strategy introduced in Section 4.2 is applied to make the network achieve a homogeneous equilibrium point. Furthermore, Chen et al. found that when only one node is added controller, the whole connected network may achieve the desired synchronization state[98]. The investigated network is a special case of (4-1) at $H = I_n$:

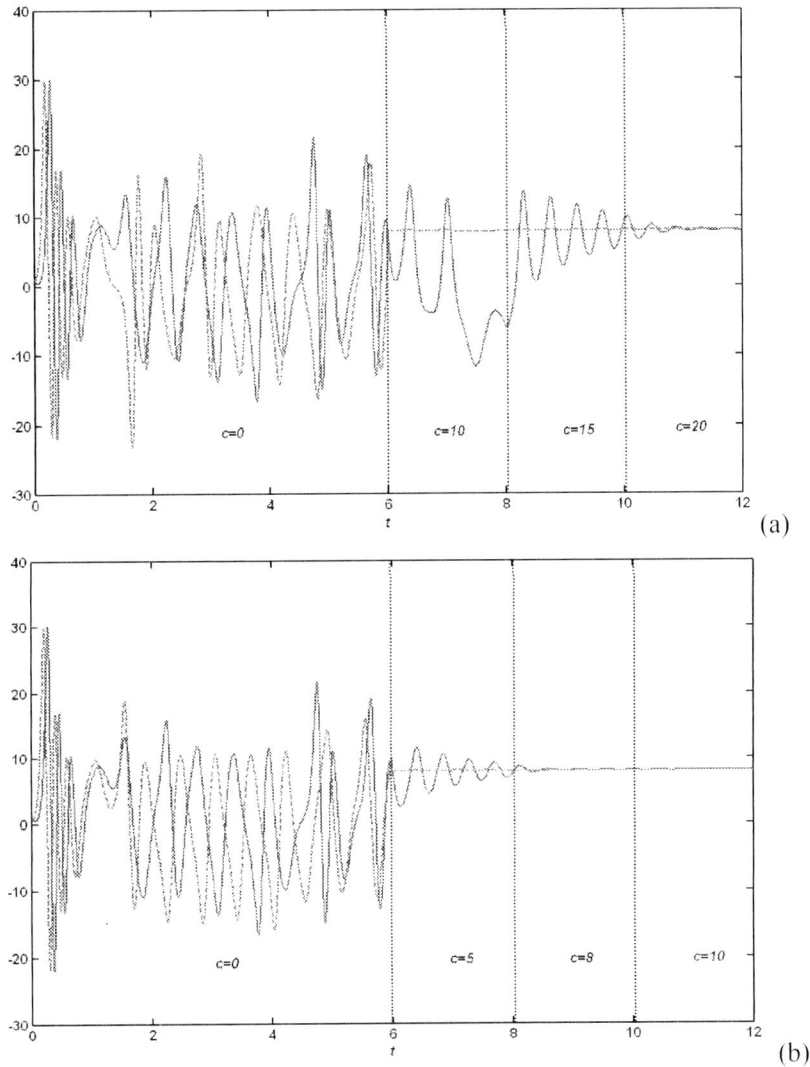

Figure 4.2. Specially pinning of scale-free network with each isolated node being Chen's system and network size $N = 50$ (a) control one node with biggest degree (b) control two nodes with biggest degrees (figure taken from [129], ©IEEE).

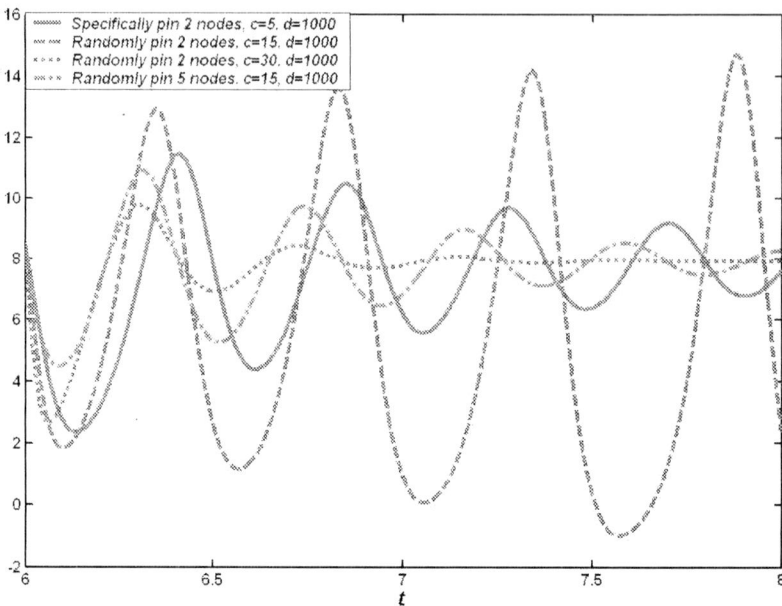

Figure 4.3. When the specially pinning strategy and random pinning strategy are adopted, the state of the node with smallest degree (figure taken from [129], ©IEEE).

$$\dot{x} = f(x_i(t),t) + c \sum_{j=1, j \neq i}^{N} a_{ij}(x_j(t) - x_i(t)), \qquad i = 1, 2, \cdots, N \qquad (4\text{-}11)$$

where the synchronization state $s(t)$ satisfies that

$$\dot{s}(t) = f(s(t),t) \qquad (4\text{-}12)$$

Without lose of generality, the first node is added controller, which is

$$\begin{cases} \dot{x}_1(t) = f(x_i(t),t) + c \sum_{j=1, j \neq i}^{N} a_{ij}(x_j(t) - x_i(t)) + c(x_1 - s(t)) \\ \dot{x}_i(t) = f(x_i(t),t) + c \sum_{j=1, j \neq i}^{N} a_{ij}(x_j(t) - x_i(t)), \qquad i = 2, \cdots, N \end{cases} \qquad (4\text{-}13)$$

Theorem 4-3[98]: Let $\mu_i(t), i = 1, \cdots, N$ be the eigenvalue of the matrix $(1/2)(Df(s(t),t) + Df^T(s(t),t))$, and $\mu(t) = \max_{i=1,\cdots,N} u_i(t)$. If $\mu(t) < -c\lambda_1 - \eta$ for all $t > 0$, where λ_1 is the minimum eigenvalue of the matrix $A - diag\{\varepsilon, 0, \cdots, 0\}$, then the network (4-11) locally stabilizes on the synchronization state $s(t)$.

Theorem 4-4[98]: If there exists positive definite diagonal matrix $P = diag\{p_1, \cdots, p_n\}$, $\Delta = diag\{\Delta_1, \cdots, \Delta_n\}$ and a constant $\eta > 0$, the inequation

$$(x-y)^T P(f(x,t) - \Delta x - f(y,t) + \Delta y)) \leq -\eta(x-y)^T(x-y) \qquad (4-14)$$

and

$$\Delta_k + c\lambda_1 < 0 \ (k = 1, \cdots, n) \qquad (4-15)$$

where λ_1 is the minimum eigenvalue of the matrix $A - diag\{\varepsilon, 0, \cdots, 0\}$, then the network (4-11) globally stabilizes on the synchronization state $s(t)$.

Theorem 4-5[98]: Under the assumptions of *Theorem 4-4*, and the updating law of coupling strength:

$$\dot{c}(t) = \frac{\alpha}{2} \sum_{i=1}^{N} (x_i(t) - s(t))^T P(x_i(t) - s(t)) \qquad (4-16)$$

where $\alpha \geq 0$ is adaptive gain, the network (4-11) globally stabilizes on the synchronization state $s(t)$.

The isolated node dynamic is also considered to be Chen's system in (4-8), the quantity $E(t) = \sqrt{\left(\sum_{i=1}^{500} |x_i(t) - s(t)|^2 / 500\right)}$ and the adaptive gain is 100. Figure 4.4 shows that the small world network with network size 500 and the isolated node being Chen's system may reach the synchronization state $s(t)$ for $t > 13s$, while at the same the coupling strength $c(t)$ is 39.63 for $t > 13s$.

The pinning control approach of adding only one controller is extended to the case of that the network has different inner coupling matrix[133]. Moreover, the pinning control strategy is applied to the adaptive complex dynamical network[134]. In addition to add controllers to few nodes, the update of each node in the network depends only on the node's neighbor states. For the network with more general topologies, the adaptive pinning control approach is also proofed to be effective[135].

4.4. CONTROLLABILITY OF PINNING CONTROL

Sorrentinal et al. found that the controllability of pinning control is relevant to the coupling strength between nodes, the control gain, and the number of pinned control [136]. Consider the dynamical state of the isolated node is

$$\dot{x}_i = f(x_i) - c\sum_{j=1}^{N} L_{ij} h(x_j) + c\sum_{k=1}^{N} \delta(i - c_k) u_i \qquad (4\text{-}17)$$

where c is the coupling strength between nodes, the definition of L_{ij} is the same as that in (1-18). $\delta(i - c_k)$ equals to 1 only if $i = c_k$; otherwise, $\delta(i - c_k)$ equals 0, which is used to select the pinned node. The control input $u_i = \kappa_i(s(t) - x_i(t))$, where κ_i is the control input of node i, the definition of $s(t)$ is in (4-12).

Define an extended $(N+1) \times (N+1)$ matrix

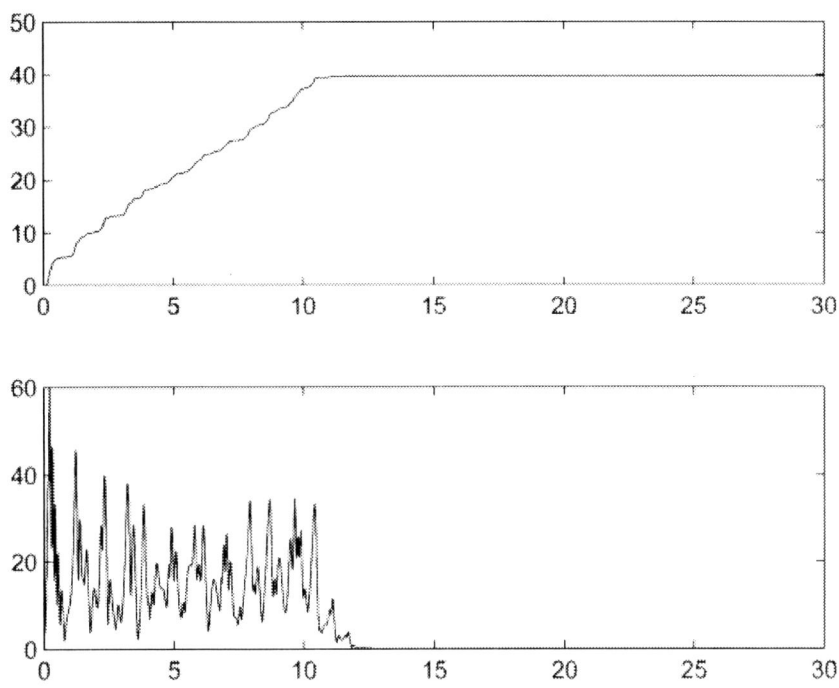

Figure 4.4. Pin 500 Chen's system coupled by a small-world network with one controller, the top is $c(t)$, the bottom is $E(t)$ (figure taken from [98], ©IEEE).

$$M = \begin{pmatrix} L_{11}+\xi_1\kappa_1 & L_{12} & \cdots & L_{1N} & -\xi_1\kappa_1 \\ L_{21} & L_{22}+\xi_2\kappa_2 & & L_{2N} & -\xi_2\kappa_2 \\ \vdots & & \ddots & & \vdots \\ L_{N1} & L_{N2} & \cdots & L_{NN}+\xi_N\kappa_N & -\xi_N\kappa_N \\ 0 & 0 & \cdots & 0 & 0 \end{pmatrix}$$

Obviously, the matrix M is asymmetric and the sum of row entries equals to 0. When the Laplace matrix L is symmetric, the matrix M may be diagonalized. Furthermore, the real of its eigenvalues may be written: $0 = u_1^r < u_2^r \leq \cdots \leq u_{N+1}^r$. Define $R^{N+1} = u_{N+1}^r / u_2^r$ and $M^{N+1} = \max_j u_j^i$, the smaller the value of R^{N+1} and M^{N+1}, the higher the controllability of the pinning control[136].

4.5. CONTROL A NETWORK TO A HETEROGENEOUS EQUILIBRIUM POINT

The pinning control of a complex network to a homogeneous point or synchronization sate may be used to complete one object. However, it is often to complete several objects for a group of connected nodes. For example, a few robots rescue several persons in the fire, where the network topology represents the communication roads between these robots. A novel approach is proposed to make the network achieve a heterogeneous equilibrium point [137]. This implies that when the whole nodes are divided into several subgroups, the nodes in the same subgroup achieve the same equilibrium point of an isolated node, while different groups correspond to different equilibrium points.

Consider a network of N diffusively linearly coupled identical oscillators, with each oscillator being an n-dimensional dynamical system. The state equations of the network are

$$\dot{x}_i(t) = f(x_i(t),t) + c \sum_{j=1, j \neq i}^{N} a_{ij}(x_j(t) - x_i(t)) + u_i(t), \quad i=1,2,\cdots,N,$$

(4-18)

where $u_i(t)$ is the control input added on node i. The definitions are the same in (1-6).

Suppose that the network is connected in the sense that there are no isolated clusters in the network. This implies that the coupling matrix A is symmetric and irreducible, whose eigenvalues can be ordered as $0 = \lambda_1(A) > \lambda_2(A) \geq \cdots \geq \lambda_N(A)$ [74, 82].

The dynamics of an isolated node without control can be written as

$$\dot{y} = f(y)$$

(4-19)

Each isolated node is assumed to have p equilibrium points $x_{ei} \in R^n$ ($i=1,2,\cdots,p$), which means $f(x_{ei})=0, (i=1,2,\cdots,p)$, where $p>1$. Suppose that each node in network (1) needs to stabilize on one of the p equilibrium points of an isolated node using distributed control. Without loss of generality, the whole nodes are divided into p clusters, the sets of subscripts of these clusters are $G_1=\{1,2,\cdots,N_1\}$, $G_2=\{N_1+1, N_1+2,\cdots,N_1+N_2\}$, $\cdots$, $G_p=\{N_1+\cdots+N_{p-1}+1,\cdots,N\}$, where $N_1+N_2+\cdots+N_p=N$. The nodes in the cluster G_i ($i=1,2,...,q$) need to stabilize on x_{ei}, where $1 \leq q \leq p$. The problem considered here is to design distributed control inputs $u_i(t)$ ($i=1,2,\cdots,N$) so that as $t \to +\infty$, there is

$$\begin{cases} x_1(t)=x_2(t)=\cdots=x_{N_1}(t)=x_{e1} \\ x_{N_1+1}(t)=x_{N_1+2}(t)=\cdots=x_{N_1+N_2}(t)=x_{e2} \\ \vdots \\ x_{N_1+\cdots+N_{q-1}+1}(t)=x_{N_1+\cdots+N_{q-1}+2}(t)=\cdots=x_N(t)=x_{eq} \end{cases} \quad (4\text{-}20)$$

In the following, an open-loop constant control approach is first proposed to make the point defined by (4-20) an inhomogeneous equilibrium point of the controlled network (4-18); then the feedback pinning control approach is applied to making the inhomogeneous equilibrium point asymptotically stable.

4.5.1. Open-loop Constant Control

The following constant control input is considered[137]:

$$u_i(t)=-c\sum_{j=1, j\neq i}^{N} a_{ij}(s_i - s_j), \quad i=1,2,\cdots,N. \quad (4\text{-}21)$$

Here s_i is the desired state of node i, i.e.,

$$\begin{cases} s_1 = s_2 = \cdots = s_{N_1} = x_{e1} \\ s_{N_1+1} = s_{N_1+2} = \cdots = s_{N_1+N_2} = x_{e2} \\ \vdots \\ s_{N_1+\cdots+N_{q-1}+1} = s_{N_1+\cdots+N_{q-1}+2} = \cdots = s_N = x_{eq} \end{cases} \quad (4\text{-}22)$$

Note that if node i does not connect to any nodes in other groups, then $u_i(t) = 0$. It is easy to verify that the desired state defined in (4-20) is an inhomogeneous equilibrium point of the network (4-18) with controller (4-21). Clearly, such an inhomogeneous equilibrium point may not be stable. However, the following numerical example shows that each node's state will be close to its desired state, so that the nodes in different groups can still be distinguished easily.

Suppose that each isolated node is a Lorenz oscillator described by

$$\begin{cases} \dot{y}_1 = -10y_1 + 10y_2 \\ \dot{y}_2 = 28y_1 - y_2 - y_1 y_3 \\ \dot{y}_3 = y_1 y_2 - 2.67 y_3 \end{cases} \quad (4\text{-}23)$$

The Lorenz oscillator has a double-scroll chaotic attractor, as shown in Figure 4.5 [18]. Three equilibrium points of the Lorenz oscillator are: $[8.4853, 8.4853, 27]^T$, $[-8.4853, -8.4853, 27]^T$ and $[0, 0, 0]^T$.

A community network is constructed by using the algorithm in [124]. The network has 100 nodes and sizes of the three communities are 28, 34, and 38, respectively (Figure 4.6). Obviously, the connections existed in a same community is much denser than the connections existed between different communities. Suppose that the desired states of nodes in three communities are three equilibrium points of the Lorenz oscillator, respectively. A constant controller of the form (4-21) is added to each node and we set the coupling strength $c = 50$. Starting from randomly initial states, the states of nodes in a same community will be close to but not exactly equal to their desired states (Figure 4.7).

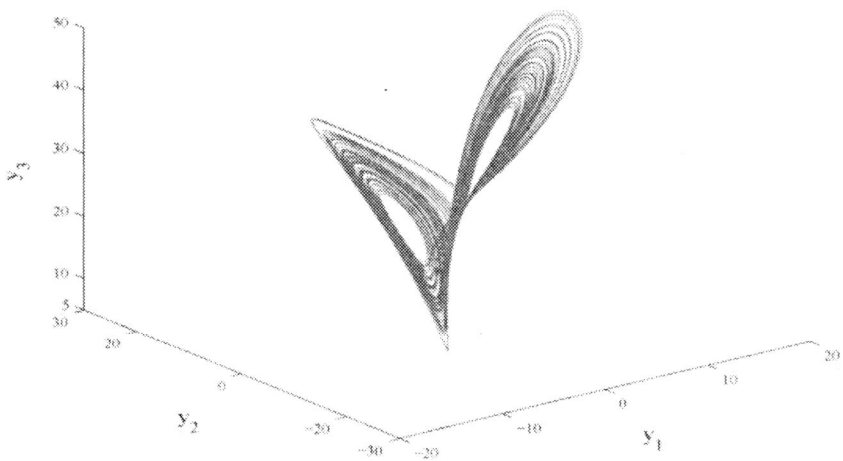

Figure 4.5. Chaotic behavior of the Lorenz oscillator(figure taken from [137], ©Elsevier).

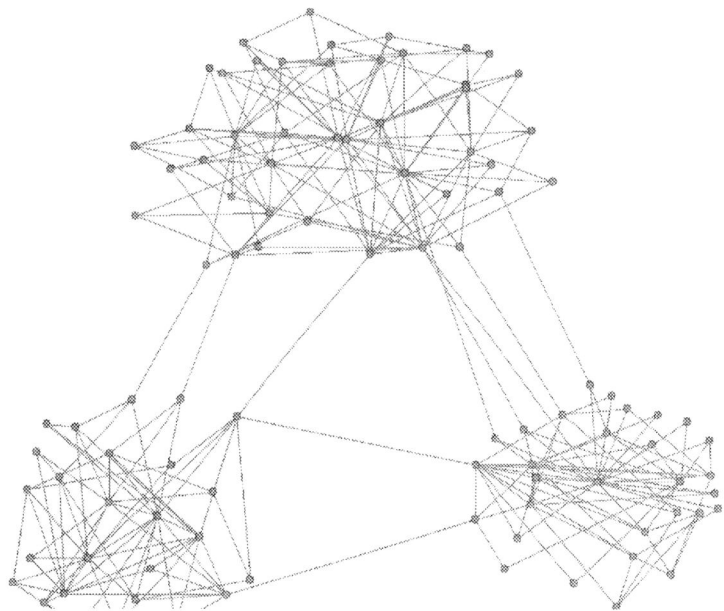

Figure 4.6. A community network model of $N = 100$ nodes with $q = 3$ communities(figure taken from [137], ©Elsevier).

4.5.2. Feedback Pinning Control

In order to stabilize the inhomogeneous equilibrium point defined in (4-20), a feedback pinning control term is added to the controller (4-21) as follows[137]:

$$u_i(t) = -c \sum_{j=1, j \neq i}^{N} a_{ij}(s_i - s_j) - h_i c d(x_i(t) - s_i) \qquad (4\text{-}24)$$

Here s_i is defined in (4-22) and $d > 0$,

$$h_i = \begin{cases} 1 & \begin{array}{l} 1 \leq i \leq \delta_{c1}, N_1 + 1 \leq i \leq N_1 + \delta_{c2}, \cdots, \\ N_1 + N_2 + \cdots + N_{q-1} + 1 \leq i \leq N_1 + N_2 + \cdots + N_{q-1} + \delta_{cq}, \end{array} \\ 0 & otherwise \end{cases} \qquad (4\text{-}25)$$

4.5.2.1. Local Stability Analysis

Denote the matrix $B = (b_{ij})_{N \times N} = A - diag[h_1, h_2, \cdots, h_N]d$. According to the proposition 1 in Ref. [98], it is easy find that the matrix B is semi-negative definite and its eigenvlues can be ordered as $0 \geq \lambda_1(B) \geq \lambda_2(B) \geq \cdots \geq \lambda_N(B)$.

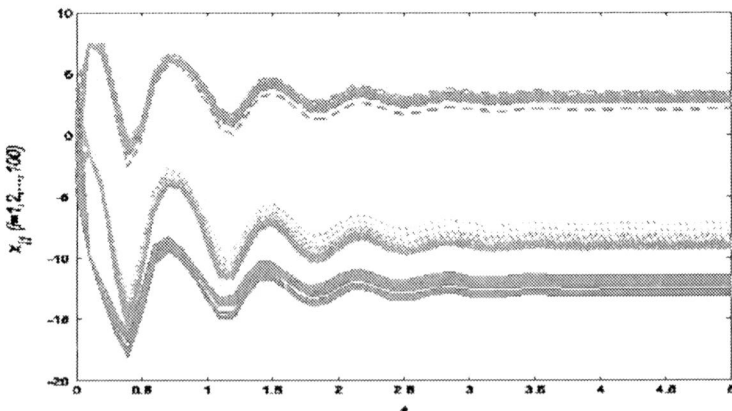

Figure 4.7 (Continued)

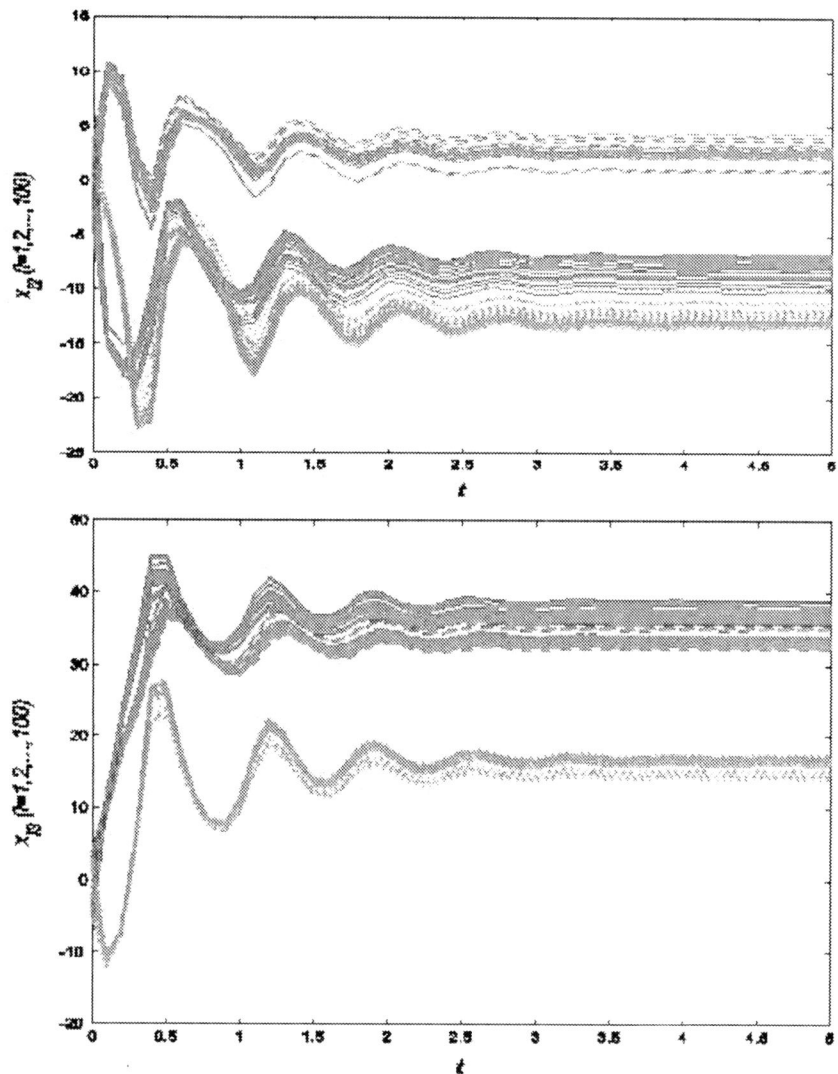

Figure 4.7. Dynamics of an open-loop controlled community network of 100 Lorenz oscillators: $x_{i1}(t)$ (top), $x_{i2}(t)$ (middle) and $x_{i3}(t)$ (bottom) (figure taken from [137], ©Elsevier).

where $1 \leq \delta_{ci} \ll N_i, 1 \leq i \leq q$. In other words, linear error feedback controllers are only added to δ_{ci} nodes in group i.

Theorem 4-6[137]. Let $\mu_i, i=1,\cdots,N$ be the eigenvalues of the matrix $\frac{1}{2}(Df(s_i)+Df^T(s_i))$, $\mu = \max_{i=1,\cdots,N} \mu_i$. If $\mu < -c\lambda_1(B)$, the point defined in (4-20) is a locally exponentially stable equilibrium point of the network (4-18) with controller (4-24).

Proof: Denote $\delta x_i(t) = x_i(t) - s_i$. It is known that

$$\frac{d\delta x_i(t)}{dt} = f(x_i(t),t) + c\left(-h_i d\delta x_i + \sum_{j=1}^{N} a_{ij}\delta x_j(t)\right), \quad i=1,\cdots,N, \quad (4\text{-}26)$$

which can be rewritten in the form

$$\frac{d\delta x_i(t)}{dt} = f(x_i(t),t) + c\sum_{j=1}^{N} b_{ij}\delta x_j(t), \quad i=1,\cdots,N. \quad (4\text{-}27)$$

Denote $\delta x(t) = [\delta x_1(t),\cdots,\delta x_N(t)] \in R^{n \times N}$. Linearization of the equation (4-17) around the origin leads to the following equation:

$$\frac{d\delta x(t)}{dt} = \left[Df(s_1)\delta x_1(t), Df(s_2)\delta x_2(t),\cdots, Df(s_N)\delta x_N(t)\right] + c\sum_{j=1}^{N} \delta x(t)B^T,$$

(4-28)

where $Df(s_i)$ is the Jacobin matrix of f at s_i.

Let $B^T = WJW^T$ be the eigenvalue decomposition of matrix B, where $J = diag[\lambda_1(B),\cdots,\lambda_N(B)]$ and $W^TW = I$. Then the equation (4-28) can be rewritten as

$$\frac{d\delta x(t)}{dt} = \left[Df(s_1)\delta x_1(t), Df(s_2)\delta x_2(t),\cdots, Df(s_N)\delta x_N(t)\right] + c\sum_{j=1}^{N} \delta x(t)WJW^T,$$

The above equation is right multiplied by W, then

$$\frac{d\delta x(t)}{dt}W = [Df(s_1)\delta x_1(t), Df(s_2)\delta x_2(t), \cdots, Df(s_N)\delta x_N(t)]W$$

$$+ c\sum_{j=1}^{N}\delta x(t)WJW^TW,$$

For $W^TW = I$, then

$$\frac{d\delta x(t)}{dt}W = [Df(s_1)\delta x_1(t), Df(s_2)\delta x_2(t), \cdots, Df(s_N)\delta x_N(t)]W$$

$$+ c\sum_{j=1}^{N}\delta x(t)W diag[\lambda_1(B), \cdots, \lambda_N(B)],$$

Denote $\delta y(t) = \delta x(t)W$. Then, it is obtained that

$$\frac{d\delta y_k(t)}{dt} = (Df(s_k) + c\lambda_k(B)I)\delta y_k(t), \qquad k = 1, \cdots, N. \qquad (4\text{-}29)$$

Obviously, it is can be seen that

$$\frac{1}{2}\frac{d\{\delta y_k^T(t)\delta y_k(t)\}}{dt} = \delta y_k^T(t)[Df(s_k) + c\lambda_k(B)I]\delta y_k(t)$$

$$= \delta y_k^T(t)[\frac{1}{2}(Df(s_k) + Df^T(s_k)) + c\lambda_k(B)I]\delta y_k(t).$$

If $\mu < -c\lambda_1(B)$, it is obviously

$$\frac{1}{2}\frac{d\{\delta y_k^T(t)\delta y_k(t)\}}{dt} \leq \delta y_k^T(t)[\mu + c\lambda_1(B)]\delta y_k(t) < 0, \qquad (4\text{-}30)$$

which means that $\delta y_k^T(t)\delta y_k(t) \leq \exp((\mu + c\lambda_1(B))t)\delta y_k^T(0)\delta y_k(0)$. The proof is complete.

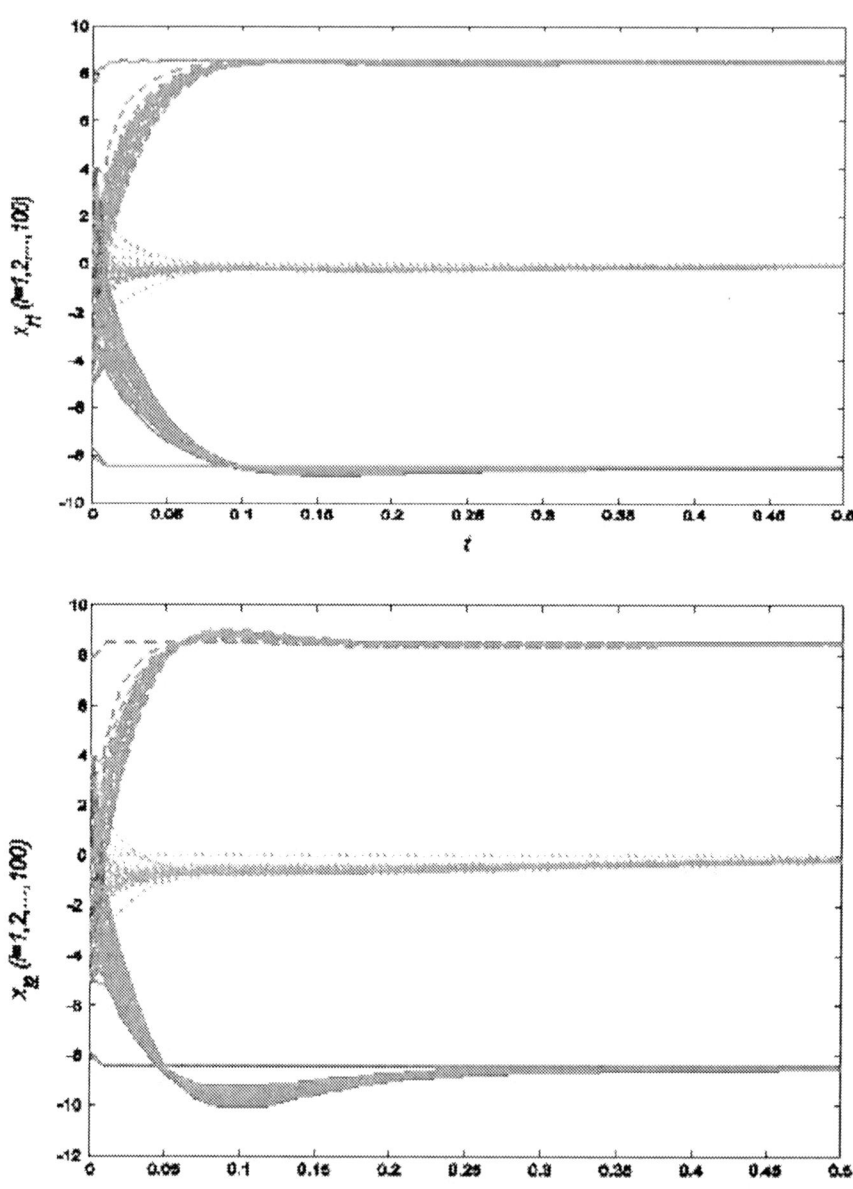

Figure 4.8 (Continued)

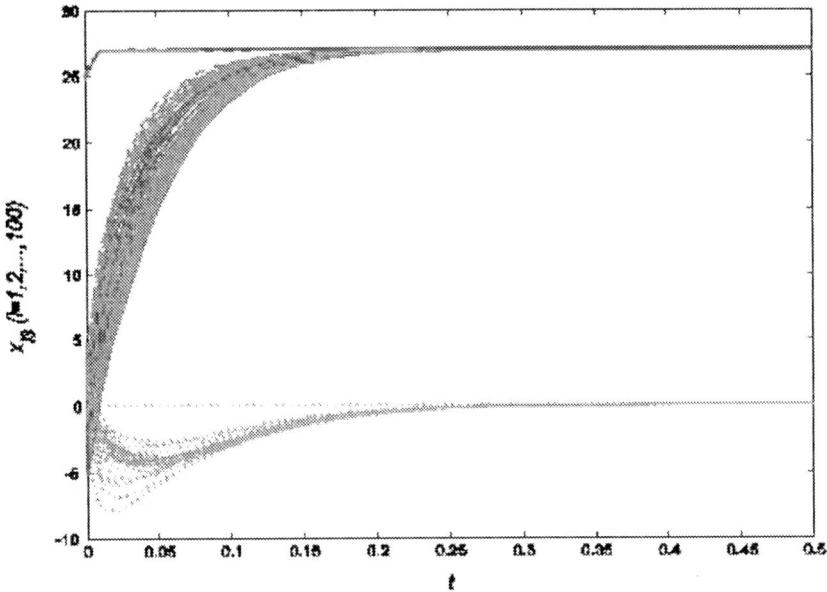

Figure 4.8. Random pinning two nodes in each community with $c = 50$ and $d = 2500$: $x_{i1}(t)$ (top), $x_{i2}(t)$ (middle) and $x_{i3}(t)$ (bottom) (figure taken from [137], ©Elsevier).

4.5.2.2. Global Stability Analysis

Theorem 4-7[137]. Suppose that there exist positive definite matrices $P = diag[p_1, \cdots, p_n]$, $\Delta = diag[\Delta_1, \cdots, \Delta_n]$ and a constant $\eta > 0$, such that

$$(x-y)^T P(f(x,t) - f(y,t) + \Delta y - \Delta x) \leq -\eta(x-y)^T(x-y) \ \forall x, y \in R^n,$$
(4-31)

and $\Delta_{max} + c\lambda_1(B) < 0$, where $\Delta_{max} = \max_{k=1,\cdots,n} \Delta_k$. Then the point defined in (4-20) is a globally exponentially stable equilibrium point of the network (4-18) with controller (4-24).

Proof: Define a Lyapunov function as

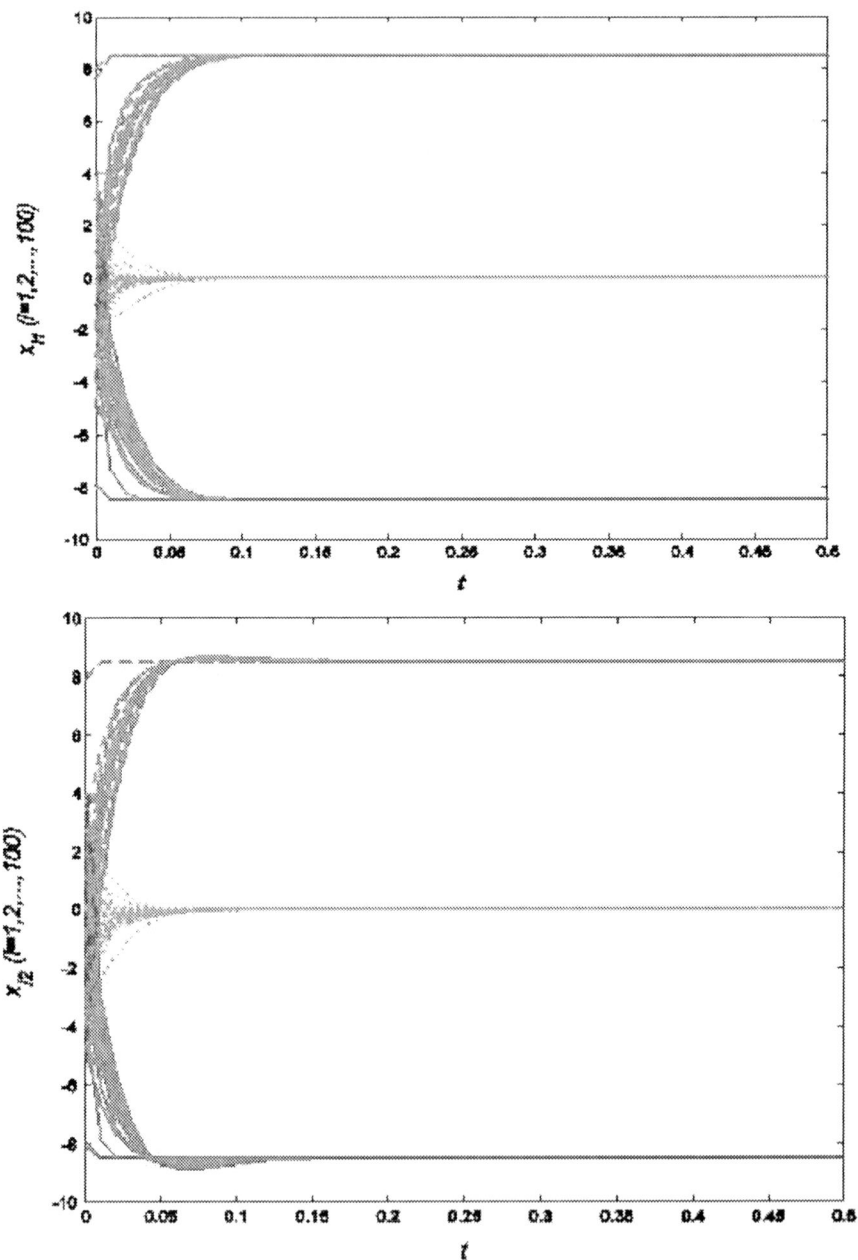

Figure 4.9 (Continued)

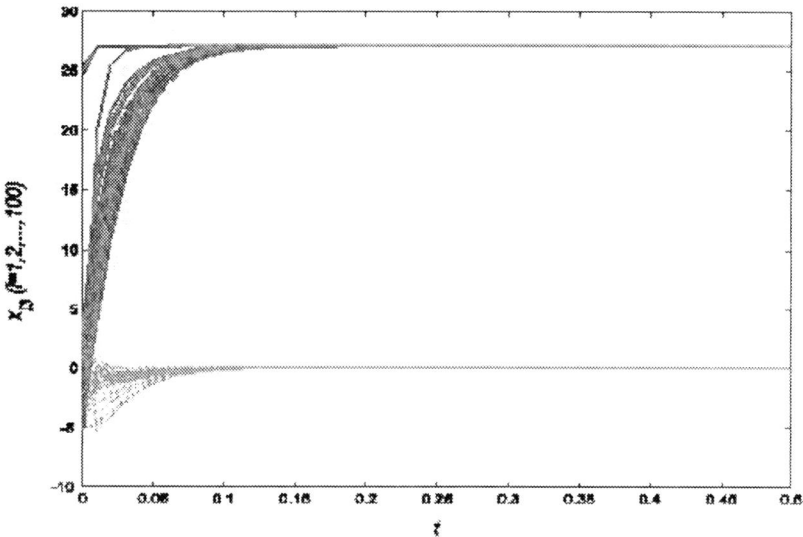

Figure 4.9. Special pinning two hub nodes in each community: $x_{i1}(t)$ (top), $x_{i2}(t)$ (middle) and $x_{i3}(t)$ (bottom) (figure taken from [137], ©Elsevier).

$$V(t) = \frac{1}{2}\sum_{i=1}^{N} \delta x_i(t)^T P \delta x_i(t). \tag{4-32}$$

Denote $\tilde{x}^k(t) = [\delta x_{1k}(t), \cdots, \delta x_{Nk}(t)]^T$, $k = 1, \cdots n$. Then, it is known that

$$\begin{aligned}
\frac{dV(t)}{dt} &= \sum_{i=1}^{N} \delta x_i(t)^T P \frac{d\delta x_i(t)}{dt} \\
&= \sum_{i=1}^{N} \delta x_i(t)^T P[f(x_i(t),t) - f(s_i) + c\sum_{j=1}^{N} b_{ij} \delta x_j(t)] \\
&= \sum_{i=1}^{N} \delta x_i(t)^T P[f(x_i(t),t) - f(s_i) - \Delta \delta x_i(t)] \\
&\quad + \sum_{i=1}^{N} \delta x_i(t)^T P[c\sum_{j=1}^{N} b_{ij} \delta x_j(t) + \Delta \delta x_i(t)]
\end{aligned}$$

$$\leq -\eta \sum_{i=1}^{N} \delta x_i(t)^T \delta x_i(t) + \sum_{i=1}^{N} \delta x_i(t)^T P[c\sum_{j=1}^{N} b_{ij} \delta x_j(t) + \Delta \delta x_i(t)]$$

$$= -\eta \sum_{i=1}^{N} \delta x_i(t)^T \delta x_i(t) + \sum_{k=1}^{n} p_k \tilde{x}^k(t)^T (cB + \Delta_k I) \tilde{x}^k(t)$$

Since $\Delta_{\max} + c\lambda_1(B) < 0$ and $0 > \lambda_1(B) \geq \cdots \geq \lambda_N(B)$, the matrix $cB + \Delta_k I$ is negative definite. it is obtained that

$$\frac{dV(t)}{dt} \leq -\eta \sum_{i=1}^{N} \delta x_i(t)^T \delta x_i(t) \leq -\eta \frac{V(t)}{\max_i p_i}. \qquad (4\text{-}33)$$

Therefore,

$$V(t) \leq \exp\left\{\frac{-\eta t}{\max_i p_i}\right\} V(0). \qquad (4\text{-}34)$$

The proof is complete.

In the special case $q = 1$, the point defined in (4-20) becomes a homogeneous equilibrium point of the controlled network. Furthermore, if $\delta_{c1} = 1$, Theorem 4-6 and Theorem 4-7 reduce to the corresponding results in [98], which showed that stabilization of a homogeneous equilibrium point can be achieved by just control one node randomly selected from the network.

4.5.2.3. Simulation Results

When the dynamical behavior of an isolated node is the chaotic Lorenz oscillator defined in (4-23), direct calculations shows that (4-31) satisfies, where $P = I_3$, $\Delta = 8I_3$, $\eta = 0.54$.

The community network shown in Figure 4.6 is considered again. Suppose that the desired states of nodes in three communities are still three equilibrium points of the Lorenz oscillator, respectively. Randomly select two nodes from each community to add linear error feedback terms. These nodes' degrees are 19, 3, 5, 3, 5, and 3, respectively. When the coupling strength is $c = 50$ and gain is $d = 2500$, the largest eigenvalue of matrix B is $\lambda_1(B) = -0.2711$. Since (4-31) holds and $\Delta_{max} + c\lambda_1(B) < 0$, state of the randomly pinning controlled network globally converges to the desired inhomogeneous equilibrium point (Figure 4.8).

Now suppose that the linear error feedback terms are added to two nodes with highest degrees in each community. The degrees of the six chosen nodes are 19, 12, 15, 17, 18 and 25, respectively. In this specific pinning case, the value of $\lambda_1(B) = -0.8435$ is much smaller than -0.2711. Comparing to the randomly pinning case, it can be seen that state of the specifically pinning controlled network converges to the desired state with a much faster speed (Figure 4.9).

Chapter 5

SYNCHRONIZATION OF TIME VARYING COMPLEX NETWORKS

ABSTRACT

In this chapter, the synchronization of complex network with switching topologies is introduced. It is proved that if the network locally synchronizes for the fixed time-average of the topology and the time-average is achieved sufficiently fast, the network locally synchronizes for switching topologies. The global synchronization of a class of directed dynamical networks with switching topologies is investigated. It is found that if there exists a directed spanning tree in the fixed time-average of network topology and the time-average is achieved sufficiently fast, then the network will reach global synchronization for appreciate coupling strength. A sufficient condition about the global synchronization is given derived. Several numerical simulations show the effectiveness of the strategy.

5.1. INTRODUCTION

Most researches have focused on the network with fixed network topology, even in Chapter 2; the investigated network has only adaptive coupling strength. However, many real world networks have time-varying coupling strength between nodes and time-varying network topology. For example, when a group of autonomous vehicles complete one subject, where each vehicle is looked as a node and communication is represented by connection between nodes. Since the limit of the communication capability and each vehicle has different velocity, the connection between them is time varying.

5.2. LOCAL SYNCHRONIZATION OF TIME VARYING COMPLEX NETWORKS

Consider a dynamical network of N linearly coupled identical oscillators, with each oscillator being an n-dimensional dynamical system. The evolution of the dynamical variables is [67, 68]

$$\dot{x}_i(t) = f(x_i(t)) + \sum_{j=1}^{N} a_{ij}(t) H(x_j(t)), \qquad (5\text{-}1)$$

where the coupling matrix $A = (a_{ij}(t)) \in R^{N \times N}$ represents the network topology: If there is a connection between node i and node j, then $a_{ij}(t) > 0$; otherwise, $a_{ij}(t) = 0$; furthermore, the dissipation condition $\sum_{j} a_{ij}(t) = 0$ satisfies. The definitions of the rest parameters are the same as those in (1-6).

Suppose the synchronization state $s(t)$ is time varying, and $\xi_i(t) = x_i(t) - s(t)$. Here if $x_1(t)$ is selected to be $s(t)$, then the network (5-1) may be rewritten

$$\dot{\xi}_i(t) = f(\xi_i(t) + s(t)) - f(s(t)) + \sum_{j=2}^{N} a_{ij}(t) H(t) \xi_j, \qquad i = 2, \cdots, N$$

(5-2)

Let $\overline{\xi}(t) = \left(\xi_2(t), \cdots, \xi_N(t)\right)^T$, then the equation (5-2) is rewritten

$$\dot{\bar{\xi}}(t) = F(t,\bar{\xi}(t)) \tag{5-3}$$

Theorem 5-1[67] Assume that $F:\Omega \to R^{n(N-1)}$ is continuously differentiable on $\Omega = \{x \in R^{n(N-1)} \mid \|x\|_2 < r\}$, with $F(t,0) = 0$ for all t, and the Jacobin matrix $DF(t,x)$ is bounded and Lipschitz on Ω, uniformly in t. Assume that there exists a bounded nonsingular real matrix $\Phi(t)$ satisfies

$$\Phi^{-1}(t)(A(t))^T \Phi(t) = diag\{\lambda_1(t), \lambda_2(t), \cdots, \lambda_N(t)\} \tag{5-4}$$

and

$$\dot{\Phi}^{-1}(t)\Phi(t) = diag\{\beta_1(t), \beta_2(t), \cdots, \beta_N(t)\} \tag{5-5}$$

Then the network (5-1) locally reaches the synchronization state $s(t)$ if and only if the linear time-varying systems

$$\dot{w} = [DF(s(t)) + \lambda_k(t)H(t) - \beta_k(t)I_n]w, \quad k = 2, \cdots, N \tag{5-6}$$

are exponentially stable about the zero solution.

The proof of this theorem may refer to the derivation of the master stability function in Section 1.3.1.

5.3. CONNECTION GRAPH STABILITY METHOD

5.3.1. Stability Analysis of Global Synchronization

Combine the Lyapunov function method and graph theory, Belykh et al. proposed a connection graph stability method to investigate the global synchronization stability of time varying networks [27, 70]. The investigated network is the same as in (5-1).

Define the difference of two nodes' states $x_{ij} = x_j - x_i, i,j = 1,\ldots,N$, then

$$\dot{x}_{ij} = f(x_j) - f(x_i) + \sum_{g=1}^{N}\left[a_{jg}Hx_{jg} - a_{ig}Hx_{ig}\right] \quad (5\text{-}7)$$

Apply the mean value theorem to the first part of the left side of (5-7), then

$$f(x_j) - f(x_i) = \left[\int_0^1 Df(\beta x_j + (1-\beta)x_i)d\beta\right]x_{ij} \quad (5\text{-}8)$$

where Df is a Jacobin matrix of f and $0 \leq \beta \leq 1$. Therefore, (5-7) may be rewritten as follows:

$$\dot{x}_{ij} = \left[\int_0^1 Df(\beta x_j + (1-\beta)x_i)d\beta\right]x_{ij} + \sum_{g=1}^{N}\left[a_{jg}Hx_{jg} - a_{ig}Hx_{ig}\right]$$

$$(5\text{-}9)$$

Adding and subtracting an additional term Bx_{ij} from (5-9), the system becomes

$$\dot{x}_{ij} = \left[\int_0^1 Df(\beta x_j + (1-\beta)x_i)d\beta - B\right]x_{ij} + BX_{ij} + \sum_{g=1}^{N}\left[a_{jg}Hx_{jg} - a_{ig}Hx_{ig}\right]$$

$$(5\text{-}10)$$

where the matrix $B = diag\{b_1,b_2,\cdots,b_n\}$. Here the nonzero diagonal entry is corresponding to that in the matrix H; $b_h \geq 0, h = 1,2,\ldots,s$; $b_h = 0, h = s+1,\ldots,n$. The object of the selection of matrix B is to make the latter two parts of the right side in (5-10) becomes stable, then the stability of (5-10) becomes the stability of the first part of the right side in (5-10).

An auditory system is designed as follows:

$$x_{ij} = \left[\int_0^1 Df(\beta x_j + (1-\beta)x_i)d\beta - B \right] x_{ij}, \quad i,j = 1,\cdots,N \quad (5\text{-}11)$$

Suppose there exists a Lyapunov function

$$W_{ij} = \frac{1}{2} x_{ij}^T E x_{ij}, \quad i,j = 1,\ldots,N \quad (5\text{-}12)$$

where the matrix $E = diag(e_1, e_2, \ldots, e_s, E_1), e_1, \ldots, e_s > 0$ and the matrix $E_1 \in R^{(N-s) \times (N-s)}$ is positive definite. In order to make the auditory system (5-11) stable, the inequation holds

$$\dot{W}_{ij} = x_{ij}^T E \left[\int_0^1 Df(\beta x_j + (1-\beta)x_i)d\beta - B \right] x_{ij} < 0 \quad (5\text{-}13)$$

In order to assure the global stability of synchronization manifold in time varying network (5-1), a Lyapunov function is constructed as follows:

$$V = \frac{1}{4} \sum_{i=1}^N \sum_{j=1}^N X_{ij}^T E x_{ij} \quad (5\text{-}14)$$

Derivative V on time t, then

$$\frac{dV}{dt} = \frac{1}{2} \sum_{i=1}^N \sum_{j=1}^N \dot{W}_{ij} + \frac{1}{2} \sum_{i=1}^N \sum_{j=1}^N X_{ij}^T B x_{ij}$$
$$- \frac{1}{2} \sum_{i=1}^N \sum_{j=1}^N \sum_{k=1}^N \left[a_{jk} x_{ji}^T EH x_{jk} + a_{ik} x_{ik}^T EH x_{ij} \right] \quad (5\text{-}15)$$

According to (5-13), the fist part of the right side in (5-15) is less than zero. In order to make the in equation $\dot{V} < 0$ hold, then

$$\frac{1}{2} \sum_{i=1}^N \sum_{j=1}^N X_{ij}^T B x_{ij} - \frac{1}{2} \sum_{i=1}^N \sum_{j=1}^N \sum_{k=1}^N \left[a_{jk} x_{ji}^T EH x_{jk} + a_{ik} x_{ik}^T EH x_{ij} \right] < 0$$

$$(5\text{-}16)$$

Furthermore, since $X_{ii}^2 = 0$ and $X_{ij}^2 = X_{ji}^2$, then the inequation (5-16) is rewritten as

$$\sum_{i=1}^{N-1}\sum_{j>i}^{N} a_{ij} x_{ij}^T EH x_{ij} > \frac{1}{N}\sum_{i=1}^{N-1}\sum_{j>i}^{N} x_{ij}^T EB x_{ij} \tag{5-17}$$

According to (5-17), the following two theorems are obtained.

Theorem 5-2[27]: Under the assumption on the eventual dissipativeness of the individual oscillator system and the assumption (5-13), the synchronization manifold of the system (5-1) is globally asymptotically stable if the following inequality holds

$$\sum_{k=1}^{m} a_{i_k j_k} x_{i_k j_k}^2 > \frac{b}{N}\sum_{i=1}^{N-1}\sum_{j>i}^{N} x_{ij}^2 \tag{5-18}$$

where m is the number of edges in the network.

The theorem 5-2 only gives a condition of the network reaching global synchronization. However, since the existence of variables x_{ij} and $x_{i_k j_k}$, it is difficult to obtain the critical coupling strength. In order to eliminate these variables, let $\tilde{x}_k = x_{i_k j_k}, k=1,\ldots,m$. For each pair nodes (i,j), let P_{ij} represent the path from node i to node j and $z(P_{ij})$ be the number of all the paths between these two nodes. Suppose one path between node i to node j crosses node $m_1, m_2, \ldots, m_v$, then $x_{ij} = x_{im_1} + x_{m_1 m_2} + \ldots + x_{m_v j}$. Therefore,

$$x_{ij}^2 = \left(\sum_{k \in P_{ij}} \pm \tilde{x}_k\right)^2 \leq z(P_{ij}) \sum_{k \in P_{ij}} \tilde{x}_k^2 \tag{5-19}$$

Combine the in-equation (5-19) and (5-18), it is easy to obtain the following theorem.

Theorem 5-3[27]: Under the assumption of Theorem 5-2, the synchronization manifold of the system (5-1) is globally asymptotically stable if

Synchronization of Time Varying Complex Networks

$$a_k(t) > \frac{b}{N} Z_k(N,m) \quad (5\text{-}20)$$

where $a_k = a_{i_k j_k}, k = 1, 2, \ldots, m$ and $Z_k(N,m) = \sum_{j>i}^{N} z(P_{ij}), k \in P_{ij}$ is the sum of the lengths of all chosen paths P_{ij} which pass through a given edge k that belongs to the coupling configuration.

For a given network, if the coupling strength of all the edges satisfies the inequation (5-20), then the network may achieve global synchronization. This implies that it is not need to calculate the Lyapunov exponent and the eigenvalue of the coupling matrix to judge the network achieving synchronization or not. When the network has irregular topology and it is difficult to obtain its eigenvalues of coupling matrix, the theorem 5-3 may be used to judge whether the network globally achieve synchronization or not.

5.3.2. Application of Connection Graph Stability Method

5.3.2.1. Average Model

Based on the nearest network model, Belykh et al. proposed a new globally coupled network model: average model [70]. In their model, each node is connected to all the other nodes, which has equal strength ε with its $2K$ nearest neighbors and has equal strength $p\varepsilon$ with the rest $N - 2K - 1$ nodes. Here $0 \le p \le 1$. The coupling strength is

$$G_{mean} = \begin{pmatrix} -g & \overset{K}{a \cdots a} & pa & \cdots & pa & \overset{K}{a \cdots a} \\ a & -g & \overset{K}{a \cdots a} & pa & \cdots pa & \overset{K-1}{a \cdots a} \\ a & a & -g & \overset{K}{a \cdots a} & pa \cdots pa & \overset{K-2}{a \cdots a} \\ & & \ddots & \ddots & \ddots & \vdots \\ \overset{K-1}{a \cdots a} & pa \cdots & pa & \overset{K}{a \cdots a} & -g & a \\ \overset{K}{a \cdots a} & pa & \cdots & pa & \overset{K}{a \cdots a} & -g \end{pmatrix} \quad (5\text{-}21)$$

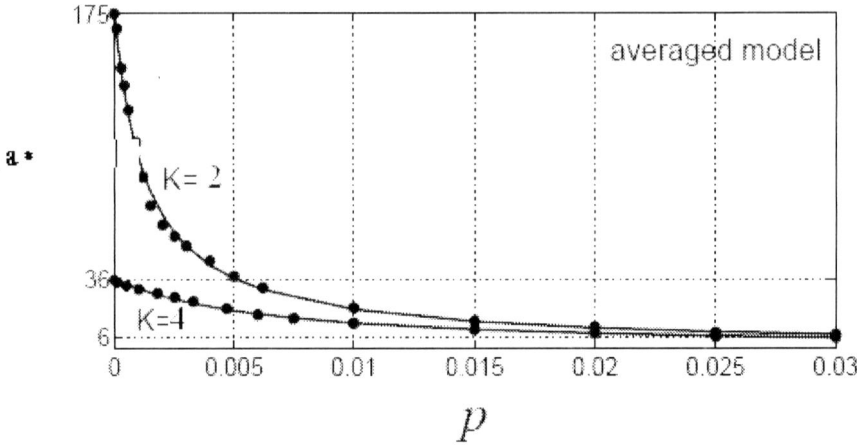

Figure 5.1. The value of synchronization threshold in the average model network with different p (figure taken from [70], ©Elsevier).

where $g = a[2K + p(N - 2K - 1)]$.

According to Theorem 5-3, the condition of the network with coupling matrix being G_{mean} is

$$a > a^* \approx \frac{b}{N} \frac{L^3(0)}{1 + p(L^3(0) - 1)} \tag{5-22}$$

Here, $L(0) = N/2K$ represents the average path length of the initial nearest network.

Figure 5.1 shows the critical coupling strength of 30 globally coupled Lorenz network with different values of p. It is easy to find that the critical coupling strength decreases rapidly when the value of p has little increase. This means that for the average model, little coupling strength decreases the synchronization threshold; furthermore, the network synchronizability is enhanced.

5.3.2.2. Blinking Small World Network

The investigated small world network model has fixed long distance edge, which means that a long distance edge is randomly selected, this edge exists forever. Recently, Belykh et al. proposed a blinking small world network, where the long distance edge exists for a while, and does not exist for a while[70]. The long distance edge selected randomly from the initial nearest network exists for a while. However, in the next short time, this long distance edge will disappear. Then, another random long distance edge will appear. In other words, during the time interval τ, each long distance edge exists according to the probability p. This process is irrelevant to the existence of other long distance edges and independence of whether the existence of the edge in the last time interval.

Consider the time varying network (5-1), where the inner coupling matrix is $H = diag(1,0,\ldots,0)$. The coupling matrix is $A = (a_{ij}(t))$. For the nearest short distance edges, that is for $|j-i| \bmod N \leq K$, $a_{ij}(t) = a_{ji}(t) = c > 0$; for the rest node pairs (i,j) in time interval $(q-1)\tau \leq t < q\tau$, $a_{ij}(t) = a_{ji}(t) = cS_{ij}(q)$. Here $S_{ij}(q)$ is a random variable, which is 1 according to the probability p, and is 0 according to the probability $1-p$.

When the random variable $S_{ij}(q)$ is independent, a theorem is obtained.

Theorem 5-4[70]: For the dynamical network (5-1), suppose each isolated system has a compact invariant set I, all the solutions of the state equation will fall in this set. In this set, suppose there exists a definite matrix E holds

$$\dot{W}_{ij} = x_{ij}^T E \left[\int_0^1 Df(\beta x_j + (1-\beta)x_i)d\beta - B \right] x_{ij} \leq \frac{1}{T_{sta}} x_{ij}^T E x_{ij}$$

(5-23)

where the parameter T_{sta} represents the characteristic synchronization time constant of the isolated system. Furthermore, the coupling strength satisfies the condition in (5-18). If the switching time τ is small enough, and the following in-equation holds

$$\left(\frac{\alpha_{exp}}{\alpha_{contr}} - 1 \right) \frac{N^2}{2} e^{-(p\gamma^2/2)(T_{syn}/\tau)(1/NcT_{stab})} < 1$$

(5-24)

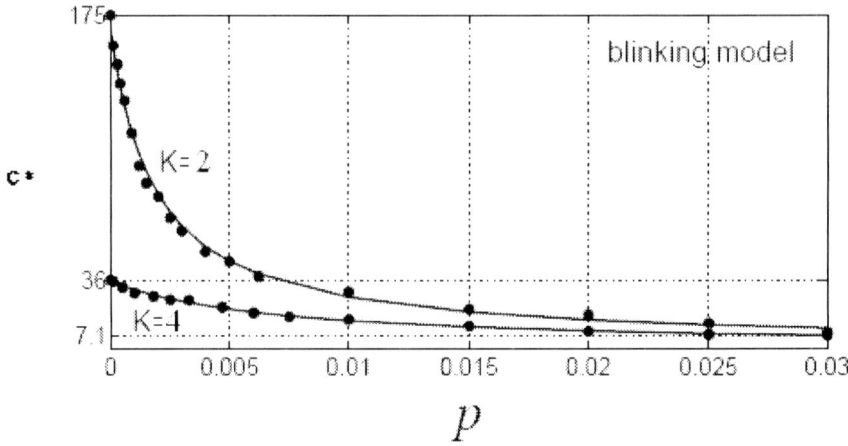

Figure 5.2. The value of synchronization threshold in the blinking model network with different p (figure taken from [70], ©Elsevier)..

Then near all the switching sequence may achieve global synchronization. Here

$$\alpha_{contr} = -\ln\left(1 - \frac{T_{syn}}{T_{stab}}\left(1 - e^{-1/ncT_{stab}}\right)\right)$$

$$\alpha_{exp} = \frac{1}{ncT_{stab}} \quad \gamma = \frac{p}{c}\left(c - \frac{\psi}{N}Z(N,K,p)\right)$$

T_{syn} represents the characteristic synchronization time constant of the coupled system. Moreover, when the in-equation holds,

$$c > c^* = \frac{\psi}{N} \cdot \frac{L^3(0)}{1 + p(L^3(0) - 1)} \tag{5-25}$$

the blinking model network may achieve global synchronization.

Compare with the average model, it is easy to find that these two models have same synchronization conditions. The only one difference is that p in the average model is a parameter; while p in the blinking model is the probability of the appearance of long distance edge. As can be seen from (5-25), the critical coupling strength decreases rapidly with the increase of the probability p. This implies that the adding of few long distance edges enhances the synchronizability of the network. Figure 5.2 shows that for the pristine nearest network with $K = 2$, when the long distance edge appear with blinking probability $p = 0.01$, the critical coupling strength decreases from 175 to 29.

5.4. FAST SWITCHING SYNCHRONIZATION OF TIME VARYING COMPLEX NETWORKS

5.4.1. Local Synchronization of Complex Networks

When a network is instantaneously disconnected, if the network could propagate sufficient information and the agents move fast enough, Skufca et al found that the moving neighborhood network may achieve complete synchronization [79]. However, it is difficult to apply the results obtained in [79], Stiwell proposed a simple version of moving neighborhood network[80].

The investigated switched system is as follows:

$$\dot{x}(t) = A_{\rho(t)} x(t) \tag{5-26}$$

where a switching sequence $\rho(t)$ is selected entries from a group of coupling matrix $\Theta = \{A_1, A_2, \cdots\}$. As we know, if each entry of Θ is Hurwitz stability and $\rho(t)$ switches slowly enough, the system (5-26) stabilizes[138]. Furthermore, if there exists a common Lyapunov function for all the coupling matrix, (5-26) stabilizes even for $\rho(t)$ is not slow [139].

When no entry of Θ is Hurwitz, Stiwell et al. found that if the switching sequence is fast enough, the system (5-26) stabilizes. Consider a network of N linearly and diffusively coupled oscillators:

$$\dot{x}_i(t) = f(x_i(t)) + c \sum_{j=1}^{N} l_{ij}(t) H x_j(t), \tag{5-27}$$

where the Laplace matrix $L = (l_{ij})_{N \times N}$ represents the interconnections between oscillators. If there is a connection between node i and node j, then $l_{ij} = -1$; otherwise, $l_{ij} = 0$. Furthermore, the diagonal entry is defined as $l_{ii} = -\sum_{j=1}^{N} l_{ij}$. The definitions of the rest parameters are the same as those in (1-6).

Suppose the synchronization state $s(t)$ satisfies

$$\dot{s}(t) = f(s(t)) \tag{5-28}$$

The subtract of (5-28) from (5-27) obtains

$$\dot{x}_i(t) - \dot{s}(t) = f(x_i(t)) - f(s(t)) + c\sum_{j=1}^{N} l_{ij}(t) H x_j(t) \tag{5-29}$$

Let $z_i(t) = x_i(t) - s(t)$, the equation (5-29) is rewritten as

$$\dot{z}_i(t) = Df(s(t)) + c\sum_{j=1}^{N} l_{ij}(t) H x_j(t) \tag{5-30}$$

where $Df(s(t))$ is the Jacobin matrix of $f(s(t))$ on t. Suppose $z(t) = \left[z_1^T, \cdots, z_N^T \right]^T$, the equation (5-30) becomes

$$\begin{aligned}\dot{z}(t) &= (I_N \otimes Df(s(t)) + c(I_n \otimes H)(L \times I_N))z(t) \\ &= (I_N \otimes Df(s(t)) + cL \otimes B)z(t) \end{aligned} \tag{5-31}$$

where $\otimes$ represents the Kronecker product.

Lemma 5-1[80]: Suppose there exists a constant T for which the matrix-valued function $E(t)$ is such that

$$\bar{E} = \frac{1}{T}\int_{t}^{t+T} E(\tau)d\tau \qquad (5\text{-}32)$$

for all t and the system

$$\dot{x}(t) = (A(t) + \bar{E})x(t), \quad x(t_0) = x_0, \quad t \geq t_0 \qquad (5\text{-}33)$$

is uniformly exponentially stable. Then there exists $\varepsilon^* > 0$ such that for all fixed $\varepsilon \in (0, \varepsilon^*)$,

$$\dot{z}(t) = (A(t) + E(t/\varepsilon))z(t), \quad z(t_0) = z_0, \quad t \geq t_0 \qquad (5\text{-}34)$$

is uniformly exponentially stable.

Based on the lemma 5-1, the following theorem is obtained:

Theorem 5-5[80]. Suppose there exists a constant T for which the matrix-valued function $L(t)$ is such that

$$\bar{L} = \frac{1}{T}\int_{t}^{t+T} L(\tau)d\tau \qquad (5\text{-}35)$$

for all t and a set of linearly coupled oscillators

$$\dot{z}_s(t) = (I_N \otimes Df(s(t)) + c\bar{L} \otimes H)z_s(t) \qquad (5\text{-}36)$$

has an asymptotically stable synchronization manifold, which means $z_i(t) = x_i(t) - s(t) \to 0$. Then there exists a positive scalar $\varepsilon^* > 0$ such that

$$\dot{z}_a(t) = (I_N \otimes Df(s(t)) + cL(t/\varepsilon) \otimes H)z_a(t) \qquad (5\text{-}37)$$

and time-varying network $L(t)$ also locally stabilize on the synchronization state $s(t)$ for all fixed $\varepsilon \in (0, \varepsilon^*)$.

Consider 25 coupled Rossler oscillators:

$$\begin{cases} \dot{x}_{i1}(t) = -x_{i2}(t) - x_{i3}(t) - c\sum_{j=1}^{25} l_{ij}(t/\varepsilon)x_{j1}(t) \\ \dot{x}_{i2}(t) = -x_{i1}(t) + 0.165x_{i2}(t) \\ \dot{x}_{i3}(t) = 0.2 + x_{i3}(t)(x_{i1}(t) - 10) \end{cases} \qquad (5\text{-}38)$$

Five graphs in Figure 5.3 are selected. Each graph is disconnected, which means that each graph has at least one isolated node. However, the union of the five graphs is connected as shown in Figure 5.3(f). The T-periodic $L(t)$ is defined as

$$L(t) = \sum_{i=1}^{5} L_i \chi_{[(i-1)T/5, iT/5]}(t) \qquad (5\text{-}39)$$

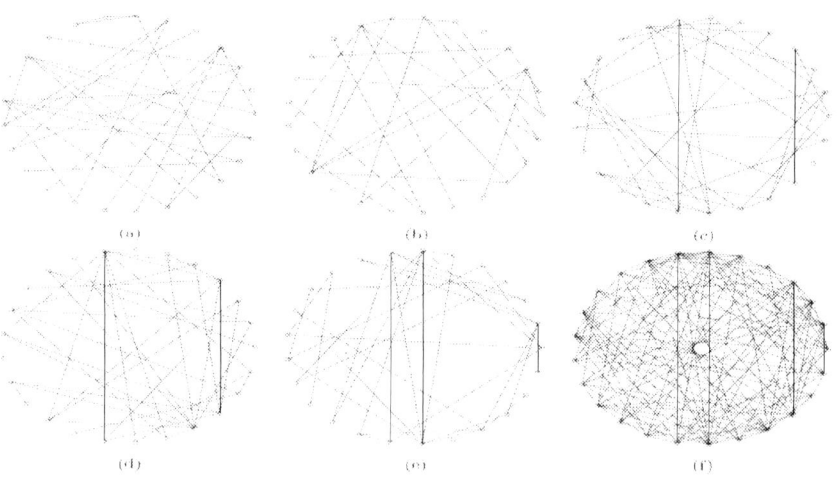

Figure 5.3. (a-e) are graphs $G_1 \sim G_5$, respectively, while (f) is the union of graphs(figure taken from [80], ©SIAM).

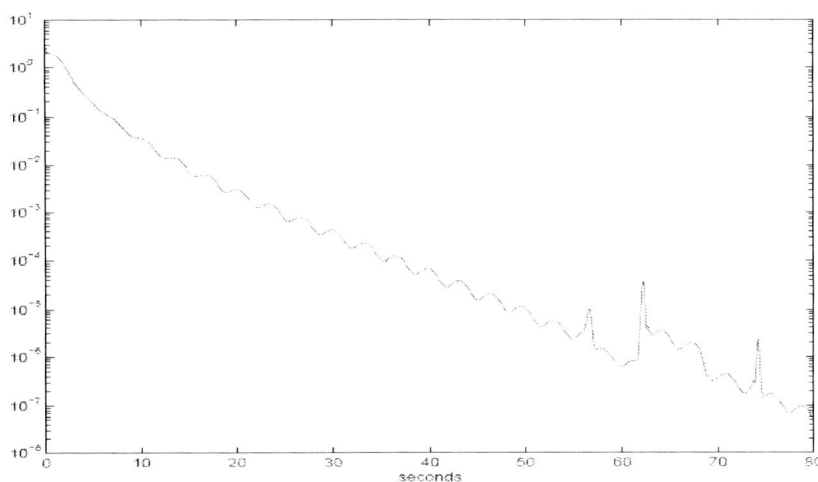

Figure 5.4. Sum-square deviation in (5-39) of 25 coupled Rossler oscillators using the average $\overline{L}$ (figure taken from [80], ©SIAM).

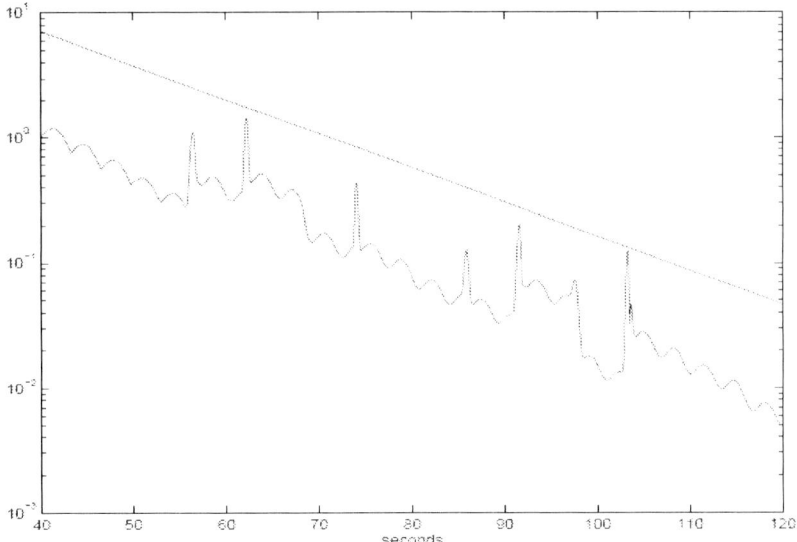

Figure 5.5. Norm of the transition matrix $\phi(t,\tau)$ along with an exponentially decaying upper bound(figure taken from [80], ©SIAM).

where $\chi_{[t_1,t_2]}(t)$ is the indicator function. The time average of $L(t)$ is $\bar{L} = \sum_{i=1}^{5} L_i / 5$. The sum-square deviation of the states is defined as

$$\sum_{i=1}^{N}(x_{i1}(t) - \frac{1}{N}\sum_{j=1}^{N}x_{j1}(t))^2 + (x_{i2}(t) - \frac{1}{N}\sum_{j=1}^{N}x_{j2}(t))^2 + (x_{i3}(t) - \frac{1}{N}\sum_{j=1}^{N}x_{j3}(t))^2$$

(5-40)

Figure 5.4 shows that the deviation is near zero when the time is bigger than 10 seconds, which implies that the network reaches synchronization. It is proofed in reference [80] that the norm is near zero when the network achieves synchronization, which is shown in Figure 5.5.

5.4.2. Global Synchronization of Directed Networks

Stiwell et al. investigated the local synchronization of networks with fast switching topologies, which means that the initial states of nodes must be near to the synchronization state [80]. However, when the initial states of nodes are randomly distributed, global synchronization of complex network is often seen in many real world networks [27, 70, 74, 82, 140, 141]. For example, it is found that a directed network globally synchronizes for sufficiently large coupling strength between nodes if there is a spanning directed tree in the network [82]. Furthermore, we investigated the global synchronization of directed networks with fast switching topologies.

Consider a network of N linearly coupled identical oscillators, with each oscillator being an n-dimensional dynamical system. The state equations of the network are

$$\dot{x}(t) = (f(x_1(t),t), \cdots, f(x_N(t),t))^T + \kappa(C(t) \otimes D(t))x(t) \qquad (5\text{-}41)$$

where $x(t) = (x_1(t), \cdots, x_N(t))^T$, $x_i(t) = [x_{i1}(t), x_{i2}(t), \cdots, x_{in}(t)]^T \in R^n$ is the state variable of node i. f describes the dynamics of each isolated node, κ is coupling strength between nodes, $D(t) = (d_{ij}(t))_{n \times n}$ is an inner coupling matrix. $C(t) = (c_{ij}(t))_{N \times N}$ is the coupling matrix which is defined as following: If there is a connection between node i and node j, then $c_{ij}(t) < 0 (i \neq j)$; otherwise, $c_{ij}(t) = 0 (i \neq j)$. $c_{ii}(t) = -\sum_{j=1, j \neq i}^{N} c_{ij}(t)$ ($i = 1, 2, \cdots, N$).

5.4.2.1. Fixed Topology

At first, the network with fixed topology is considered in [82]. This means that in (5-41) $C(t) = C$ for all t, the following lemma is obtained.

Lemma 5-2 [82]. Assume that:

(i) C is a zero row sums matrix with nonpositive off-diagonal elements.
(ii) $f(x(t), t) + \kappa D(t)x(t)$ is V-uniformly decreasing for some symmetric positive definite V, that is $(x - y)^T V(f(x,t) + \kappa D(t)x - f(y,t) - \kappa D(t)y) \leq -\mu \|x - y\|^2$ for some $\mu > 0$ and all $x, y \in R^n$ and all t.
(iii) $VD(t) \leq 0$ and is symmetric for all t.
(iv) the underlying weighted directed graph contains a spanning directed tree.

Then the network (5-41) globally synchronizes for sufficiently large coupling strength κ.

5.4.2.2. Switching Topologies

When the differences of the nodes' states are small, synchronization of networks with switching topology is investigated in [18]. It was found that if the network with fixed time-average topology can reach synchronization and the time-average is achieved sufficiently fast, the network with switching topologies will reach synchronization. However, the investigated network in [18] is undirected. For the directed network with switching topologies, a sufficient condition of the network reaching synchronization is obtained, especially despite of the differences of the nodes' states.

Theorem 5-6[142]. Suppose there exists a constant T for which the matrix-valued function $C(t)$ is such that

$$\bar{C} = \frac{1}{T}\int_t^{t+T} C(\tau)d\tau \tag{5-42}$$

for all t and the system

$$\dot{x}(t) = (f(x_1(t),t),\cdots,f(x_N(t),t))^T + (\bar{C}\otimes D(t))x(t) \tag{5-43}$$

satisfies the conditions (i)-(iv) in the Lemma. This means the system (5-40) globally synchronizes. Furthermore, suppose that the oscillator has bounded slope such that

$$\left|(f(v,t)-f(v',t))/(v-v')\right| \leq l,$$

where $l > 0$ for all $v, v' \in R^n$ and $v \neq v'$.

Then there exists $\varepsilon^* > 0$ such that for all fixed $\varepsilon \in (0,\varepsilon^*)$, the system

$$\dot{x}(t) = (f(x_1(t),t),\cdots,f(x_N(t),t))^T + (C(t/\varepsilon)\otimes D(t))x(t) \tag{5-44}$$

globally synchronizes.

Proof:
Since the system (5-43) satisfies the conditions (i-iv) in the Lemma 5-2, there exists a symmetric irreducible zero row sums matrix $U_{N\times N}$ with nonpositive off-diagonal elements. Construct a Lyapunov function

$$g(x(t)) = \frac{1}{2}x(t)^T(U\otimes V)x(t) \tag{5-45}$$

Then, it is found that

$$\dot{g}(x(t)) = x(t)^T (U \otimes V) \dot{x}(t)$$

$$= x(t)^T (U \otimes V) \begin{pmatrix} f(x_1(t), t) + D(t) x_1(t) \\ \vdots \\ f(x_N(t), t) + D(t) x_N(t) \end{pmatrix}$$

$$+ x(t)^T (U \otimes V)(\overline{C} \otimes D(t) - I \otimes D(t)) x(t)$$

$$\leq \sum_{i<j} -U_{ij} (x_i(t) - x_j(t))^T V(f(x_i(t), t) + D(t) x_i(t) - f(x_j(t), t) - D(t) x_j(t))$$

$$\leq \sum_{i<j} -U_{ij} (-\mu \| x_i(t) - x_j(t) \|^2)^T$$

Note that $-U_{ij} \geq 0$ for $i < j$. For each $-U_{ij} \geq 0$ and $\delta > 0$, and sufficiently large t such that if $\| x_i(t) - x_j(t) \| \geq \delta$, then $\dot{g} \leq -(\mu/2) \| x_i(t) - x_j(t) \|^2$. This implies that for large enough t, $\| x_i(t) - x_j(t) \| \leq \delta$. Therefore, $\lim_{t \to \infty} \| x_i(t) - x_j(t) \| = 0$.

For system (5-43), suppose $\overline{e}(t) = (\overline{e}_1(t), \cdots, \overline{e}_N(t))^T$, $\overline{e}_i(t) = x_i(t) - s(t)$, where $s(t)$ is the synchronization state $s(t) = x_i(t), i = 1, \cdots, N$, which satisfies and $\dot{s}(t) = f(s(t))$.

Then construct the following Lyapunov function as follows

$$g(\overline{e}(t)) = \frac{1}{2} \overline{e}(t)^T (U \otimes V) \overline{e}(t) \tag{5-46}$$

According to the above analysis of (5-45), it also can be obtained that $\lim_{t \to \infty} \| x_i(t) - x_j(t) \| = 0$. Then there exist positive scalars η, ρ and φ

$$\eta \| \overline{e}(t) \|^2 \leq g(\overline{e}(t), t) \leq \rho \| \overline{e}(t) \|^2 \tag{5-47}$$

$$\dot{g}(\overline{e}(t), t) \leq -\varphi \| \overline{e}(t) \|^2 \tag{5-48}$$

For, then (5-43) can be rewritten as

$$\dot{\bar{e}}_i(t) = f(x_i(t)) - f(s(t)) + \kappa \sum_{j=1}^{N} \bar{c}_{ij} D(t) \bar{e}_j(t) \quad (5\text{-}49)$$

For $e_i(t) = x_i(t) - s(t)$, (5-44) can be rewritten as

$$\dot{e}_i(t) = f(x_i(t)) - f(s(t)) + \kappa \sum_{j=1}^{N} c_{ij}(t/\varepsilon) D(t) e_j(t) \quad (5\text{-}50)$$

Because $|f(v,t) - f(v',t)/v - v'| \leq l$, without loss of generality, assume $v - v' > 0$, then

$$-l(v - v') \leq f(v,t) - f(v',t) \leq l(v - v') \quad (5\text{-}51)$$

Then (5-50) can be rewritten as

$$\dot{e}_i(t) = f(x_i(t)) - f(s(t)) + \kappa \sum_{j=1}^{N} c_{ij}(t/\varepsilon) D(t) e_j(t) \leq l(x_i(t) - s(t))$$

$$+ \kappa \sum_{j=1}^{N} c_{ij}(t/\varepsilon) D(t) e_j(t)$$

$$(5\text{-}52)$$

$$-l(x_i(t) - s(t)) + \kappa \sum_{j=1}^{N} c_{ij}(t/\varepsilon) D(t) e_j(t) \leq \dot{e}_i(t) = f(x_i(t)) - f(s(t))$$

$$+ \kappa \sum_{j=1}^{N} c_{ij}(t/\varepsilon) D(t) e_j(t)$$

$$(5\text{-}53)$$

Combined (5-49) and the right of (5-51), it is obtained

$$\dot{\bar{e}}(t) = (lI + \kappa \bar{C} \otimes D(t)) \bar{e}(t) \quad (5\text{-}54)$$

The right of (5-52) can be written

$$\dot{e}(t) = (lI + \kappa C(t/\varepsilon) \otimes D(t)) e(t) \quad (5\text{-}55)$$

where I is an identity matrix.

To establish global stability of (5-44), $g(e(t),t)$ is needed to be also a Lyapunov function for (5-44) if ε is sufficiently small. This claim is achieved by showing that for sufficiently small value of ε, $\Delta g(e(t), t + \varepsilon T, t) = g(e(t + \varepsilon T), t + \varepsilon T) - g(e(t), t)$ is negative definite for all t

$$\Delta g(e(t), t + \varepsilon T, t) = \frac{1}{2} e(t + \varepsilon T)^T (U \otimes V) e(t + \varepsilon T) - \frac{1}{2} e(t)^T (U \otimes V) e(t)$$
(5-56)

Suppose $\phi_C(t, t_0)$ is the transition matrix corresponding to $lI + \kappa C(t/\varepsilon) \otimes D(t)$, i.e.

$$e(t) = \phi_C(t, t_0) e(t_0)$$

Then (5-55) can be rewritten

$$\Delta g(e(t), t + \varepsilon T, t) = \frac{1}{2} e(t)^T (\phi_C^T(t + \varepsilon T, t)(U \otimes V)\phi_C(t + \varepsilon T, t) - (U \otimes V))e(t)$$
(5-57)

Similarly, let $\phi_{\overline{C}}(t, t_0)$ be the transition matrix corresponding to $lI + \kappa \overline{C} \otimes D(t)$, the Peano-Baker series representation of the transition matrix is used to define

$$\phi_C(t + \varepsilon T, t) = I + \int_t^{t+\varepsilon T} (lI + \kappa C(\sigma_1/\varepsilon) \otimes D(t)) d\sigma_1$$
$$+ \sum_{i=2}^{\infty} \int_t^{t+\varepsilon T} (lI + \kappa C(\sigma_1/\varepsilon) \otimes D(t)) \int_t^{\sigma_1} \cdots \int_t^{\sigma_{i-1}} (lI + \kappa C(\sigma_1/\varepsilon) \otimes D(t)) d\sigma_1 \cdots d\sigma_i$$

By hypothesis,

$$\int_t^{t+\varepsilon T} C(\sigma/\varepsilon)d\sigma = \varepsilon T \bar{C}$$

Then it is obtained

$$H(t+\varepsilon T,t) = \phi_C(t+\varepsilon T,t) - \phi_{\bar{C}}(t+\varepsilon T,t)$$
$$= \sum_{i=2}^{t}\int_t^{t+\varepsilon T}(lI+\kappa C(\sigma_1/\varepsilon)\otimes D(t))\int_t^{\sigma_1}\cdots\int_t^{\sigma_{i-1}}(lI+\kappa C(\sigma_1/\varepsilon)\otimes D(t))d\sigma_1\cdots d\sigma_i$$
$$-\sum_{i=2}^{t}\int_t^{t+\varepsilon T}(lI+\kappa\bar{C}\otimes D(t))\int_t^{\sigma_1}\cdots\int_t^{\sigma_{i-1}}(lI+\kappa\bar{C}\otimes D(t))d\sigma_1\cdots d\sigma_i$$

Suppose

$$\|lI+\kappa C(t/\varepsilon)\otimes D(t)\|\le\alpha, \|lI+\kappa\bar{C}\otimes D(t)\|\le\alpha,$$

a bound for $H(t+\varepsilon T,t)$ is computed

$$\|H(t+\varepsilon T,t)\|\le e^{2\varepsilon T\alpha}-1-2\varepsilon T\alpha \tag{5-58}$$

Note that $\phi_C(t+\varepsilon T,t) = H(t+\varepsilon T,t)+\phi_{\bar{C}}(t+\varepsilon T,t)$, then (5-57) can be rewritten

$$\Delta g(e(t),t+\varepsilon T,t) = \frac{1}{2}e(t)^T(\phi_{\bar{C}}^T(t+\varepsilon T,t)(U\otimes V)\phi_{\bar{C}}(t+\varepsilon T,t)-(U\otimes V))e(t)$$
$$+\frac{1}{2}e(t)^T(\phi_{\bar{C}}^T(t+\varepsilon T,t)(U\otimes V)H(t+\varepsilon T,t)+H^T(t+\varepsilon T,t)(U\otimes V)\phi_{\bar{C}}(t+\varepsilon T,t)$$
$$+H^T(t+\varepsilon T,t)(U\otimes V)H(t+\varepsilon T,t))e(t)$$

$$\tag{5-59}$$

It is known from (5-47) and (5-48) that

$$\|U\otimes V\|\le\rho \tag{5-60}$$

$$\|\phi_{\tilde{e}}(t,t_0)\| \le \sqrt{\rho/\eta} e^{-\frac{\varphi}{2\rho}(t-t_0)} \tag{5-61}$$

$$g(\overline{e}(t),t) \le e^{-\frac{\varphi}{\rho}(t-t_0)} g(\overline{e}(t_0),t_0) \tag{5-62}$$

From (5-46) and (5-62),

$$g(\overline{e}(t+\varepsilon T),t+\varepsilon T) - g(\overline{e}(t),t) \le (e^{-\frac{\varphi\varepsilon T}{\rho}} - 1)g(\overline{e}(t),t) \le \rho(e^{-\frac{\varphi\varepsilon T}{\rho}} - 1)\|\overline{e}(t)\|^2$$

Thus

$$g(e(t+\varepsilon T),t+\varepsilon T) - g(e(t),t) \le (e^{-\frac{\varphi\varepsilon T}{\rho}} - 1)g(e(t),t) \le \rho(e^{-\frac{\varphi\varepsilon T}{\rho}} - 1)\|e(t)\|^2$$

$$\frac{1}{2}e(t)^T(\phi_{\tilde{e}}^T(t+\varepsilon T,t)(U\otimes V)\phi_{\tilde{e}}(t+\varepsilon T,t) - (U\otimes V))e(t) \le \rho(e^{-\frac{\varphi\varepsilon T}{\rho}} - 1)\|e(t)\|^2 \tag{5-63}$$

From (5-58), (5-60), (5-61), (5-62) and (5-63)

$$\Delta g(e(t),t+\varepsilon T,t) \le (\rho(e^{-\varphi\varepsilon T/\rho} - 1) + \rho(\sqrt{\rho/\eta}e^{-\varphi\varepsilon T/2\rho})(e^{2\varepsilon T\alpha} - 1 - 2\varepsilon T\alpha)$$
$$+\frac{\rho}{2}(e^{2\varepsilon T\alpha} - 1 - 2\varepsilon T\alpha)^2)\|e(t)\|^2 \tag{5-64}$$

Defining the continuously differentiable function $q(\varepsilon,e_a(t))$ to be the right-hand side of (5-64), it can be shown that $q(0,e(t)) = 0$ and $\frac{\partial}{\partial \varepsilon}q(0,e(t)) = -\varphi T\|e(t)\|^2 < 0$. Thus since $q(\varepsilon,e(t)) \to \infty$ as $\varepsilon \to \infty$, there exists ε^*, $q(\varepsilon^*,e(t)) = 0$ and $q(\varepsilon,e(t)) < 0$ for all $\varepsilon \in (0,\varepsilon^*)$. Thus $\Delta g(e(t),t+\varepsilon T,t) \le 0$ for all $\varepsilon \in (0,\varepsilon^*)$.

Similarly, the stability of the left of (5-53) is obtained. Therefore, the stability of (5-50) is obtained. This means the system (5-44) globally synchronizes.

The proof is complete.

5.4.4.3. Simulation Results

As a typical example, the Chua's circuit is considered as each isolated node's dynamics in the complex dynamical network:

$$\begin{cases} \dot{y}_1 = 9y_2 - 2.5714 y_1 + 1.9286[|y_1+1|-|y_1-1|] \\ \dot{y}_2 = y_1 - y_2 + y_3 \\ \dot{y}_3 = -14.2857 y_2 \end{cases} \quad (5\text{-}65)$$

A set of three network topologies in Figure 5.6 (a)-(c) are selected, and each of them is disconnected. However, the union of them, as shown in Figure 5.6 (d), has a directed spanning tree with the root node being Node 9. This implies that the condition (iv) of the lemma is satisfied and there is at least one path between the root node and each other node.

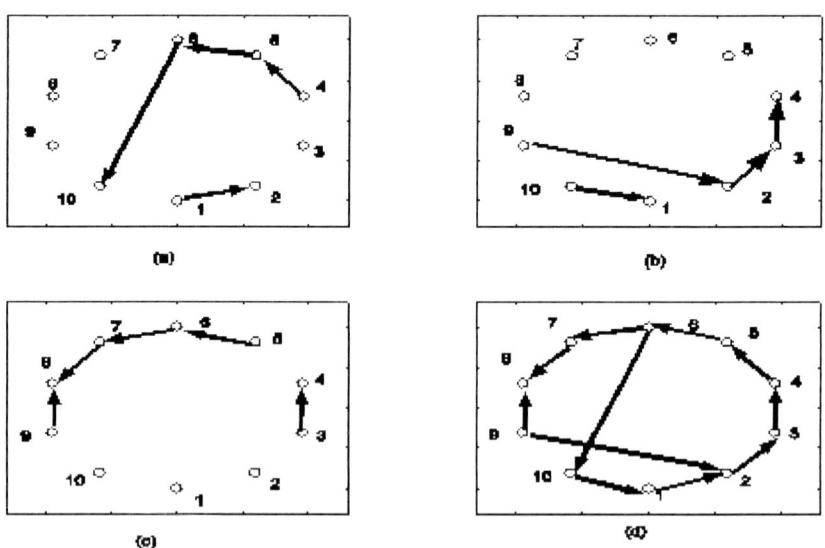

Figure 5-6. (a)-(c) are network topologies G_1 - G_3, respectively, while (d) is the union of them(figure taken from [142], ©Communications in Theoretical Physics).

The coupling matrix of G_1, G_2 and G_3 are denoted by C_1, C_2 and C_3, respectively. The weight of each edge in Figure 5.6(a)-(c) is assigned to -1. Let $\overline{C} = \frac{1}{\varepsilon T}\int_0^{\varepsilon T} C(t/\varepsilon)dt = \blacklozenge C_1 + C_2 + C_3 \blacklozenge /3$ denotes the coupling matrix of the union of C_1, C_2 and C_3. Since the matrix $C_i (i = 1,2,3)$ is a zero row sums matrix with non-positive off-diagonal elements, the matrix $\overline{C}$ is also a zero row sums matrix with non-positive off-diagonal elements. This means the condition (i) in the lemma is satisfied. For Chua circuit, when the coupling strength is $\kappa = 50$, the matrix $V = I$, and the matrix $D(t) = -I$, where I is an identity matrix. Obviously, the fixed time-average network satisfies the conditions (ii)-(iii) in the lemma. Therefore, from the lemma, the network with the coupling matrix $\overline{C}$ may reach global synchronization in theory.

Figure 5-7 shows the states of nodes in the network with the coupling matrix $\overline{C}$. As can be seen from Figure 5.7, the states of dimension x_{i1} reach synchronization after 0.7s; the states of dimension x_{i2} reaches synchronization after 0.9s; the states of dimension x_{i3} reaches synchronization after 1s. This means that for $t > 1s$, the network reach synchronization.

When the node is Chua circuit, the in-equation $|(f(v,t) - f(v',t))/(v-v')| \leq l$ satisfies with the largest slope bound being $l = 3.9421$. Combined with the network with the coupling matrix $\overline{C}$ satisfies the conditions (i)-(iv) of the lemma, it is found that the network with switching topologies G_1, G_2 and G_3 may globally reach synchronization when the switch time is sufficiently small according to the theorem.

Figure 5-8 shows that the dynamics of network with switching topologies G_1, G_2 and G_3, where the coupling strength is $\kappa = 50$ and the switching time is $\varepsilon = 0.003$ s. It is found that the first state of dimension x_{i1} reaches synchronization after 0.6s; the states of dimension x_{i2} reaches synchronization after 1s; the states of dimension x_{i3} reaches synchronization after 1s. This means that for $t > 1s$, the network reaches synchronization. The whole ten nodes can reach global synchronization after 1s.

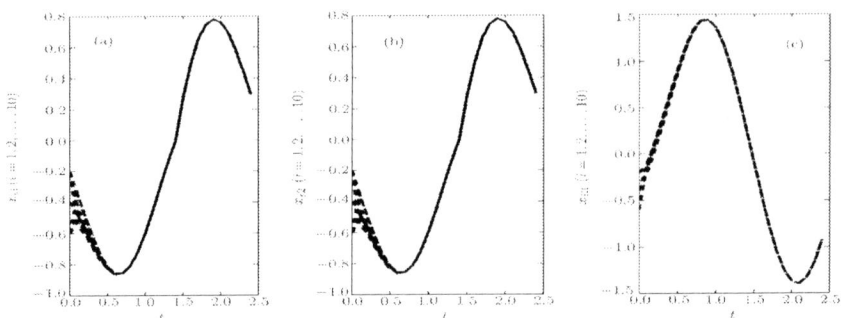

Figure 5.7. Dynamics of the fixed network in Figure 5.6(d) and each node being an isolated Chua's circuit and coupling strength $\kappa = 50$. (a) $x_{i1}(t)$; (b) $x_{i2}(t)$; (c) $x_{i3}(t)$ (figure taken from [142], ©Communications in Theoretical Physics) ..

Figure 5-9 shows that the dynamics of the same network as used in Figure 5.6. Here, the switching time increases to $\varepsilon = 0.03$ s. It is found that the states of all three dimensions can not reach synchronization. This implies that different topologies must be switched fast to make the network achieve synchronization.

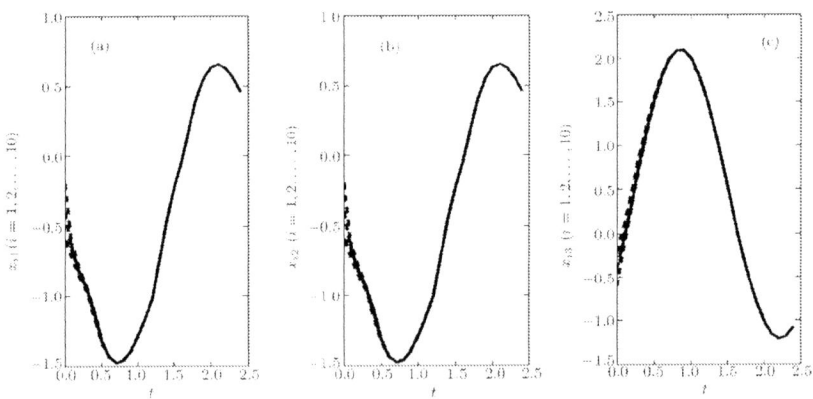

Figure 5.8. Dynamics of the network with switching topologies in Figure 5.6(a)-(c) and each node being an isolated Chua's circuit. Coupling strength $\kappa = 50$ and switching time $\varepsilon = 0.003$ s. (a) $x_{i1}(t)$; (b) $x_{i2}(t)$; (c) $x_{i3}(t)$ (figure taken from [142], ©Communications in Theoretical Physics) .

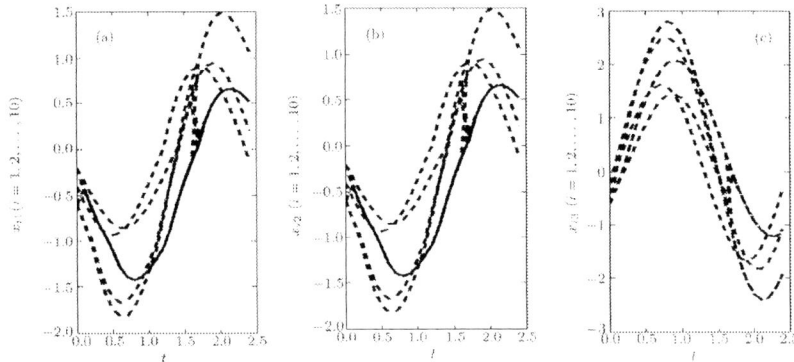

Figure 5.9. Dynamics of the network with switching topologies in Figure 5.6(a)-(c) and each node being an isolated Chua's circuit. The coupling strength $\kappa = 50$ and switching time $\varepsilon = 0.03$ s. (a) $x_{i1}(t)$; (b) $x_{i2}(t)$; (c) $x_{i3}(t)$ (figure taken from [142], ©Communications in Theoretical Physics).

ACKNOWLEDGMENTS

Xin Biao Lu acknowledges financial support from the Fundamental Research Funds for the Central Universities under Grant No. 2009B20114. Bu Zhi Qin acknowledges financial support from the Natural Science Foundation for Young Scholars of Nanjing College of Chemical Technology.

REFERENCES

[1] Huygenii, *Horoloquim oscilatoriumm*. France,Parisiism, 1673.
[2] Strogatz, S.H. and Stewart, I., Coupled oscillators and biological synchronization. *Scientific. American*, 1993. 269:102-109.
[3] Neda, Z., et al., The sound of many hands clapping. *Nature*, 2000. 403(6772):849.
[4] Neda, Z., et al., Physics of the rhythmic applause. *Physical Review E - Statistical Physics, Plasmas, Fluids, and Related Interdisciplinary Topics*, 2000. 61(6 B):6987.
[5] Gray, C.M., Synchronous oscillations in neuronal systems: Mechanisms and functions. *Journal of Computational Neuroscience*, 1994. 1:11-38.
[6] Glass, L., Synchronization and rhythmic processes in physiology. *Nature*, 2001. 410:277-284.
[7] Mohanty, P., Nanotechnology: Nano-oscillators get it together. *Nature*, 2005. 437(7057):325.
[8] Kaka, S., et al., Mutual phase-locking of microwave spin torque nano-oscillators. *Nature*, 2005. 437(7057):389.
[9] A. Pikovsky, M.R., J. Kurths, *Synchronization*. 2001, Cambridge,UK: Cambridge University Press.
[10] Osipov G.V., K.J., Zhou C., *Synchronization in oscillatory networks*. 2007, Berlin, Germany: Springer.
[11] Winfree, A.T., Biological rhythms and the behavior of populations of coupled oscillators. *Journal of Theoretical Biology*, 1967. 16(1):15.
[12] Winfree, A.T., *The Geometry of Biological Time*. 1980, Berlin, Germany: Springer-Verlag.

[13] Ariaratnam J.T., S.S.H., Phase diagram for the winfree model of coupled nonlinear oscillators. *Physics Review Letter*, 2001. 86:4278-4281.
[14] Watts, D.J. and Strogatz, S.H., Collective dynamics of "small-world" networks. *Nature*, 1998. 393:440-442.
[15] Barabasi, A.-L. and Albert, R., Emergence of scaling in random networks. *Science*, 1999. 286:509-512.
[16] Strogatz, S.H., Exploring complex networks. *Nature*, 2001. 410(6825):268.
[17] Albert, R. and Barabasi, A.L., Statistical mechanics of complex networks. *Reviews of Modern Physics*, 2002. 74(1):47.
[18] Dorogotsev, S.N.a.M., J. F. F., *Evolution of Networks:* From Biological Nets to the Internet and WWW. 2003.
[19] Newman, M.E.J., The structure and function of complex networks. *SIAM Review*, 2003. 45(2):167-256.
[20] Boccaletti S., L.V., Moreno Y., Chavez M., Hwang D.-U., Complex networks: Structure and dynamics. *Physics Report*, 2006. 424:175-308.
[21] Arenas A, G.A., Kurths J., Moreno Y., Zhou C. S., synchronization in complex networks. *Physics Report*, 2008. 469:93-153.
[22] Newman, M.E.J., The structure and function of networks. *Computer Physics Communications*, 2002. 147(1-2):40.
[23] Newman, M.E.J., Assortative mixing in networks. *Physical Review Letters*, 2002. 89(20):2087011.
[24] Arenas, A., et al., Synchronization in complex networks. *Physics Reports-Review Section of Physics Letters*, 2008. 469(3):93-153.
[25] Barahona, M. and Pecora, L.M., Synchronization in small-world Systems. *Physical Review Letters*, 2002. 89(5):054101.
[26] Chen, Y., Rangarajan, G., and Ding, M., General stability analysis of synchronized dynamics in coupled systems. *Physical Review E - Statistical, Nonlinear, and Soft Matter Physics*, 2003. 67(2 2):262091.
[27] Belykh, V.N., Belykh, I.V., and Hasler, M., Connection graph stability method for synchronized coupled chaotic systems. *Physica D: Nonlinear Phenomena*, 2004. 195(1-2):159-187.
[28] Lu, X.B. and Qin, B.Z., Transition from non-synchronization to synchronization of complex networks. *Physica a-Statistical Mechanics and Its Applications*, 2009. 388(24):5024-5028.
[29] Lu, X.B., Li, X., and Wang, X.F., Synchronization in triangled complex networks. *Communications in Theoretical Physics*, 2006. 45(5):955-960.

[30] Vasalou, C., Herzog, E.D., and Henson, M.A., Small-World Network Models of Intercellular Coupling Predict Enhanced Synchronization in the Suprachiasmatic Nucleus. *Journal of Biological Rhythms*, 2009. 24(3):243-254.

[31] Motter, A.E., Zhou, C., and Kurths, J., Network synchronization, diffusion, and the paradox of heterogeneity. *Physical Review E - Statistical, Nonlinear, and Soft Matter Physics*, 2005. 71:016116.

[32] Pecora, L.M., Synchronization conditions and desynchronizing patterns in coupled limit-cycle and chaotic systems. *Physical Review E - Statistical Physics, Plasmas, Fluids, and Related Interdisciplinary Topics*, 1998. 58(1):347.

[33] Fink, K.S., et al., Three coupled oscillators as a universal probe of synchronization stability in coupled oscillator arrays. *Physical Review E - Statistical Physics, Plasmas, Fluids, and Related Interdisciplinary Topics*, 2000. 61(5 A):5080.

[34] Lu, X.B., Wang, X.F., Li, X., and Fang, J.Q., Synchronization in weighted complex networks: Heterogeneity and synchronizability. *Physica a-Statistical Mechanics and Its Applications*, 2006. 370(2):381-389.

[35] Lu, X.B., Wang, X.F., and Fang, J.Q., Topological transition features and synchronizability of a weighted hybrid preferential network. *Physica a-Statistical Mechanics and Its Applications*, 2006. 371(2):841-850.

[36] Pecora, L.M. and Carroll, T.L., Master stability functions for synchronized coupled systems. *Physical Review Letters*, 1998. 80:2109.

[37] Heagy, J.F., Carroll, T.L., and Pecora, L.M., Synchronous chaos in coupled oscillator systems. *Physical Review E*, 1994. 50(3):1874-1885.

[38] Heagy, J.F., Pecora, L.M., and Carroll, T.L., Short wavelength bifurcations and size instabilities in coupled oscillator systems. *Physical Review Letters*, 1995. 74(21):4185.

[39] Ding, M. and Yang, W., Stability of synchronous chaos and on-off intermittency in coupled map lattices. *Physical Review E - Statistical Physics, Plasmas, Fluids, and Related Interdisciplinary Topics*, 1997. 56(4):4009.

[40] Rangarajan, G. and Ding, M., Stability of synchronized chaos in coupled dynamical systems. *Physics Letters A*, 2002. 296:204-209.

[41] Chavez, M., Hwang, D.U., Martinerie, J., and Boccaletti, S., Degree mixing and the enhancement of synchronization in complex weighted networks. *Physical Review E*, 2006. 74(6):066107.

[42] Kocarev, L. and Amato, P., *Synchronization in power-law networks. Chaos,* 2005. 15(2):024101.
[43] Wang, X.F. and Chen, G., Synchronization in Scale-free Dynamical Networks: Robustness and Fragility. *IEEE Transactions on Circuits and Systems-I: Fundamental Theory and Applications,* 2002. 49:54-62.
[44] Li, X. and Chen, G., Synchronization and Desynchronization of Complex Dynamical Networks: An Engineering Viewpoint. *IEEE Transactions on Circuits and Systems I: Fundamental Theory and Applications,* 2003. 50(11):1381-1390.
[45] Wang, X.F. and Chen, G., *Synchronization in small-world dynamical networks.* International Journal of Bifurcation & Chaos, 2002. 12:187-192.
[46] Liu, C.e.a., Analyzing and controlling the network synchronization regions. *Physica a-Statistical Mechanics and Its Applications,* 2007. 386:531-542.
[47] Wu, C.W. and Chua, L.O., Synchronization in an array of linearly coupled dynamical systems. *IEEE Transactions on Circuits and Systems I: Fundamental Theory and Applications,* 1995. 42(8):430.
[48] Heagy, J.F., Carroll, T.L., and Pecora, L.M., *Desynchronization by periodic orbits. Physcal Review E,* 1995. 52(2):1253-1256.
[49] Chua, L.O., Yang, T., Zhong, G.-Q., and Wu, C.W., Synchronization of Chua's circuits with time-varying channels and parameters. *IEEE Transactions on Circuits and Systems I: Fundamental Theory and Applications,* 1996. 43(10):862.
[50] Wang, X.F., Complex networks: topology, dynamics and synchronization. *International Journal of Bifurcation and Chaos,* 2002. 12:885-916.
[51] Nishikawa, T., Motter, A.E., Lai, Y.C., and Hoppensteadt, F.C., Heterogeneity in oscillator networks: Are smaller worlds easier to synchronize? *Physical Review Letters,* 2003. 91(1):014101/1.
[52] Duan Z S, C.G.R., and Huang L., Complex network synchronizability: analysis and control. *Physical Review E - Statistical Physics, Plasmas, Fluids, and Related Interdisciplinary Topics,* 2007. 76:056103.
[53] Fan, J., Li, X., and Fan Wang, X., On synchronous preference of complex dynamical networks. *Physica A: Statistical Mechanics and its Applications,* 2005. 355(2-4):657.
[54] Fan, J., Wang, X.F., and Li, X., Synchronization-Optimal Rewiring of Power Law Networks. *Physical Review E,* 2006.

[55] Donetti, L., Hurtado, P.I., and Munoz, M.A., Entangled networks, synchronization, and optimal network topology. *Physical Review Letters*, 2005. 95(18):188701.
[56] Wang X F, L.X., Chen G R, *Complex Network Theory and Its Application*, ed. T.U. Press. 2006.
[57] Korniss, G., Synchronization in weighted uncorrelated complex networks in a noisy environment: Optimization and connections with transport efficiency. *Physical Review E*, 2007. 75(5):051121.
[58] Wang, X.G., Lai, Y.C., and Lai, C.H., Enhancing synchronization based on complex gradient networks. *Physical Review E*, 2007. 75(5):056205.
[59] Toroczkai, Z. and Bassler, K.E., Jamming is limited in scale-free systems. *Nature*, 2004. 428(6984):716-716.
[60] Park, K.H., Lai, Y.C., Zhao, L., and Ye, N., Jamming in complex gradient networks. *Physical Review E*, 2005. 71(6):065105.
[61] Hwang, D.U., Chavez, M., Amann, A., and Boccaletti, S., Synchronization in complex networks with age ordering. *Physical Review letters*, 2005. 94(13):138701.
[62] Zou, Y.L., Zhu, J., and Chen, G.R., Synchronizability of weighted aging scale-free networks. *Physical Review E*, 2006. 74(4):046107.
[63] Chavez, M., et al., Synchronization is enhanced in weighted complex networks. *Physical Review Letters*, 2005. 94:218701.
[64] Nishikawa, T. and Motter, A.E., Synchronization is optimal in nondiagonalizable networks. *Phyiscs Review E*, 2006. 73:065106.
[65] Li, Z. and Lee, J.J., New eigenvalue based approach to synchronization in asymmetrically coupled networks. *Chaos*, 2007. 17:043117.
[66] Li, C. and Chen, G., Synchronization in general complex dynamical networks with coupling delays. *Physica A: Statistical Mechanics and its Applications*, 2004. 343(1-4):263.
[67] Lu, J., Yu, X., and Chen, G., Chaos synchronization of general complex dynamical networks. *Physica A: Statistical Mechanics and its Applications*, 2004. 334(1-2):281.
[68] Lu, J. and Chen, G., A time-varying complex dynamical network model and its controlled synchronization criteria. *IEEE Transactions on Automatic Control*, 2005. 50(6):841-846.
[69] Lu, J., Yu, X., Chen, G., and Cheng, D., Characterizing the synchronizability of small-world dynamical networks. *IEEE Transactions on Circuits and Systems I: Regular Papers*, 2004. 51(4):787.

[70] Belykh, I.V., Belykh, V.N., and Hasler, M., Blinking model and synchronization in small-world networks with a time-varying coupling. *Physica D: Nonlinear Phenomena*, 2004. 195(1-2):188-206.

[71] Hauptmann, C., et al., Control of spatially patterned synchrony with multisite dealyed feedback. *Physical Review E*, 2007(76(6)):066209.

[72] Senthilkumar, D.V. and Lakshmanan, M., Intermittency transtion to generalized synchronization in coupled time-delay systems. *Physical Review E*, 2007. 76(6):066210.

[73] Hong, H., Kim, B.J., Choi, M.Y., and Park, H., Factors that predict better synchronizability on complex networks. *Physical Review E - Statistical, Nonlinear, and Soft Matter Physics*, 2004. 69(6 2):067105.

[74] Wu, C.W., Perturbation of coupling matrices and its effect on the synchronizability in arrays of coupled chaotic systems. *Physics Letters, Section A: General, Atomic and Solid State Physics*, 2003. 319(5-6):495.

[75] Zhou, C., Motter, A.E., and Kurths, J., Universality in the synchronization of weighted random networks. *Physical Review Letters*, 2006. 96(3).

[76] Motter, A.E., Zhou, C.S., and Kurths, J., Enhancing complex-network synchronization. *Europhysics Letters*, 2005. 69(3):334.

[77] Atay, F.M., Jost, J., and Wende, A., Delays, connection topology, and synchronization of coupled chaotic maps. *Physical Review Letters*, 2004. 92(14):144101.

[78] Barabasi, A.L., Albert, R., and Jeong, H., Mean-field theory for scale-free random networks. *Physica A: Statistical Mechanics and its Applications*, 1999. 272(1):173.

[79] Skufca, J.D. and Bollt, E.M., Commnication and synchronization in disconnected networks with dynamic topology: moving neighborhood networks. *Mathematical biosciences and engineering*, 2004. 1(2):347-359.

[80] Stilwell, D.J., Bollt, E.M., and Roberson, D.G., Sufficient conditions for fast switching synchronization in time-varying network topologies. *SIAM Journal on Applied Dynamical Systems*, 2006. 5(1):140.

[81] Ito, J. and Kaneko, K., Spontaneous structure formation in a network of chaotic units with variable connection strengths. *Physics Review Letters*, 2002. 88:028701.

[82] Wu, C.W., Synchronization in arrays of coupled nonlinear systems with delay and nonreciprocal time-varying coupling. *IEEE Transactions on Circuits and Systems II-Express Briefs*, 2005. 52(5):282-286.

[83] Stanley, K., Bryant, B. D., Mikkulainen, R, Evolving adaptive neural networks with and without adaptive synapses. *Proceedings of the 2003 IEEE Congress on Evolutionary Computation*, 2003.
[84] Fewell, J.H., Social Insect Network. *Science*, 2003. 301:1867-1870.
[85] Chen, G. and Li, Z., Robust adaptive synchronization of uncertain dynamical networks. *Physics Letters A*, 2004. 324:166-178.
[86] Zhou, C.S. and Kurths, J., Dynamical weights and enhanced synchronization in adaptive complex networks. *Physical Review Letters*, 2006. 96:164102.
[87] Lu, J.A., Lü, J., and Zhou, J., Adaptive synchronization of an uncertain complex dynamical network. *IEEE Trans. Automat. Control*, 2006. 51(4):652-656.
[88] Moallemi, C.C. and Van, R.B., Distributed optimization in adaptive networks. *Advances in Neural Information Systems*, ed. S. Thrun, L.K. Saul, and B. Schölkopf. 2004, Cambridge: MIT Press. 887-894.
[89] Newman, M.E.J., Barabàsi, A.L., and Watts, D.J., *The Structure and Dynamics of Complex Networks*. 2006: Princeton University Press.
[90] Ren, Q.S., Yu, X.Q., Li, T.T., and Zhao, J.Y., Using an adaptive scheme to reduce the coupling cost in chaotic phase synchronization of complex networks. *Europhysics letters*, 2008. 83(2):20002.
[91] Ren, W., Beard, R.W., and Atkins, E.M., Information consensus in multivehicle cooperative control. *IEEE Control System Magazine*, 2007. 27:71-82.
[92] De Lellis, P., di Bernardo, M., and Garofalo, F., Synchronization of complex networks through local adaptive coupling. *Chaos*, 2008. 18(3):037110.
[93] De Lellis, P., Di Bernardo, M., Sorrentino, F., and Tierno, A., Adaptive synchronization of complex networks. *International Journal of Computer Mathematics*, 2008. 85(8):1189-1218.
[94] De Lellis, P., Di Bernardo, M., and Garofalo, F., Novel decentralized adaptive strategies for the synchronization of complex networks. *Automatica*, 2009. 45(5):1312-1318.
[95] Lu, W.L., Adaptive sychronization networks via neighborhood information: synchronization and pinning control. *Chaos*, 2007. 17:023122.
[96] Zhou, J., Lu, J.A., and Lu, J.H., Adaptive synchronization of an uncertain complex dynamical network. *IEEE Transaction on Automatic Control*, 2006. 51(4):652-656.

[97] Li, D.M., Lu, J.A., Wu, X.Q., and Chen, G., Estimating the bounds for the Lorenz family of chaotic systems. *Chaos,Solitons Fractals*, 2005. 23:529-534.

[98] Chen, T.P., Liu, X., and Lu, W.L., Pinning complex networks by a single controller. *Ieee Transactions on Circuits and Systems I-Regular Papers*, 2007. 54(6):1317-1326.

[99] Kacperski, K. and Holyst, J.A., Opinion formation model with strong leader and external impact: a mean field approach. *Physica A*, 1999. 269(2-4):511-526.

[100] Zachary, W.W., An information flow model for conflict and fission in small groups. *Journal of anthropological research*, 1977. 33:452-473.

[101] Yoshioka, M., Cluster synchronization in an ensemble of neurons interacting through chemical synapses. *Physical Review E*, 2005. 71(6):061914.

[102] Holme, P., Huss, M., and Jeong, H.W., Subnetwork hierarchies of biochemical pathways. *Bioinformatics*, 2003. 19(4):532-538.

[103] Sporns, O., Chialvo, D.R., Kaiser, M., and Hilgetag, C.C., Organization, development and function of complex brain networks. *Trends in Cognitive Sciences*, 2004. 8(9):418-425.

[104] Flake, G.W., Lawrence, S., Giles, C.L., and Coetzee, F.M., Self-organization and identification of web communities. *Computer*, 2002. 35(3):66-71.

[105] Rulkov, N.F., Images of synchronized chaos: experiments with circuits. *Chaos*, 1996. 6:262-279.

[106] Kaneko, K., Relevance of dynamic clustering to biological networks. *Physica D-Nonlinear Phenomena*, 1994. 75:55-73.

[107] Belykh, V., Belykh, I., and Mosekilde, E., Cluster synchronization modes in an ensemble of coupled chaotic oscillators. *Physics Review E.*, 2001. 63:036216.

[108] Belykh, I., Belykh, V., Nevidin, K., and Hasler, M., Persistent clusters in lattices of coupled nonidentical chaotic systems. *Chaos.* 13(1):165-178.

[109] Popovych, O., Pikovsky, A., and Maistrenko, Y., Cluster-splitting bifurcation in a system of coupled maps. *Physica D-Nonlinear Phenomena*, 2002. 168:106-125.

[110] Qin, W.X. and Chen, G.R., Coupling schemes for cluster synchronization in coupled Josephson equations. *Physica D*, 2004. 197:375-391.

[111] De Smet, F. and Aeyels, D., Clustering in a network of non-identical and mutually interacting agents. *Proceedings of the Royal Society a-Mathematical Physical and Engineering Sciences*, 2009. 465(2103):745-768.
[112] McGraw, P.N. and Menzinger, M., Clustering and the synchronization of oscillator networks. *Physical Review E - Statistical, Nonlinear, and Soft Matter Physics*, 2005. 72:015101.
[113] Newman, M.E.J., Clustering and preferential attachment in growing networks. *Physical Review E*, 2001. 64(2):025102.
[114] Wu, W., Zhou, W.J., and Chen, T.P., Cluster Synchronization of Linearly Coupled Complex Networks Under Pinning Control. *Ieee Transactions on Circuits and Systems I-Regular Papers*, 2009. 56(4):829-839.
[115] Ma, Z.J., Zhang, G., Wang, Y., and Liu, Z.R., Cluster synchronization in star-like complex networks. *Journal of Physics a-Mathematical and Theoretical*, 2008. 41(15):155101.
[116] Wang, K.H., Fu, X.C., and Li, K.Z., Cluster synchronization in community networks with nonidentical nodes. *Chaos*, 2009. 19(2):023106.
[117] Cao, J.D. and Li, L.L., Cluster synchronization in an array of hybrid coupled neural networks with delay. *Neural Networks*, 2009. 22(4):335-342.
[118] Sinatra, R., et al., Cluster Structure of Functional Networks Estimated from High-Resolution Eeg Data. *International Journal of Bifurcation and Chaos*, 2009. 19(2):665-676.
[119] Lu, X.B., Qin, B.Z., and Lu, X.Y., New approach to cluster synchronization in complex dynamical networks. *Communications in Theoretical Physics*, 2009. 51(3):485-489.
[120] Lu, X.B. and Qin, B.Z., Adaptive cluster synchronization in complex dynamical networks. *Physics Letters A*, 2009. 373(40):3650-3658.
[121] Ma, Z.J. and Liu, Z.R., A new method to realize cluster synchronization in connected networks. *Chaos*, 2006. 16:023103.
[122] Lu, X.B. and Qin, B.Z. Adaptive cluster synchronization in coupled phase oscillators. in *International Conference on Information Engineering and Computer Science*. 2009. Wuhan, China.
[123] Belykh, V., et al., Cluster synchronization in oscillatory networks. *Chaos*, 2008. 18:037106.

[124] Li, C.G. and Chen, G.R., Modelling of weighted evolving networks with community structures. *Physica a-Statistical Mechanics and Its Applications*, 2006. 370(2):869-876.
[125] Zhou, C.S. and Kurths, J., Hierarchical synchronization in complex networks with heterogeneous degrees. *Chaos*, 2006. 16:015104.
[126] Hu, G. and Qu, Z.L., Controlling spatiotemporal chaos in coupled map lattice systems. *Physical Review Letters*, 1994. 72(1):68-71.
[127] Roy, R., Murphy, T.W., Maier, J.T.D., and Gills, Z., Dynamical control of a chaotic laser: Experimental stabilization of a globally coupled system. *Physical Review Letters*, 1992. 68:1259-1262.
[128] Wang, X.F. and Chen, G.R., Pinning control of scale-free dynamical networks. *Physica a-Statistical Mechanics and Its Applications*, 2002. 310(3-4):521-531.
[129] Li, X., Wang, X.F., and Chen, G.R., Pinning a complex dynamical network to its equilibrium. *Ieee Transactions on Circuits and Systems I-Regular Papers*, 2004. 51(10):2074-2087.
[130] Lu, W.L. and Chen, T.P., New approach to synchronization analysis of linearly coupled ordinary differential systems. *Physica D-Nonlinear Phenomena*, 2006. 213(2):214-230.
[131] Xiang, L.Y., et al., Pinning control of complex dynamical networks with general topology. *Physica a-Statistical Mechanics and Its Applications*, 2007. 379(1):298-306.
[132] Liu, Z.X., Chen, Z.Q., and Yuan, Z.Z., Pinning control of weighted general complex dynamical networks with time delay. *Physica a-Statistical Mechanics and Its Applications*, 2007. 375(1):345-354.
[133] Zhan, M., Gao, J., Wu, Y., and Xiao, J., Chaos synchronization in coupled systems by appling pinning control. *Phyiscs Review E*, 2007. 76:036203.
[134] Lu, W.L., Adaptive dynamical networks via neighborhood information: Synchronization and pinning control. *Chaos*, 2007. 17(2):23122.
[135] Zhou, J., Lu, J.A., and Lu, J.H., Pinning adaptive synchronization of a general complex dynamical network. *Automatica*, 2008. 44:996-1003.
[136] Sorrentino, F., di Bernardo, M., Garofalo, F., and Chen, G.R., Controllability of complex networks via pinning. *Physical Review E*, 2007. 75(4):046103.
[137] Lu, X.B., Wang, X.F., and Fang, J.Q., Controlling a complex dynamical network to attain an inhomogeneous equilibrium. *Physica D-Nonlinear Phenomena*, 2010. 239(7):341-347.

[138] Liberzon, D. and Morse, A.S., Basic problems in stability and design of switched systems. *IEEE Transaction on Control Systems,* 1999. 5(19):59-70.

[139] Liberzon, D., *Switching in systems and control.* 2003, Boston: Birkhauser.

[140] Wu, J.S. and Jiao, L.C., Global synchronization and state tuning in asymmetric complex dynamical networks. *IEEE Transactions on Circuits and Systems II-Express Briefs*, 2008. 55(9):932-936.

[141] Lu, X.B. and Z., Q.B. Adaptive global synchronization of directed networks with fast switching technologies. in *Proceedings of the 8th world congress on intelligent control and authomation.* 2010. Ji Nan, China.

[142] Lu, X.B. and Qin, B.Z., Global synchronization of directed networks with fast switching topologies. *Commun. Theor. Phys.,* 2009(52):1019-1024.

INDEX

A

algorithm, 20, 53, 80
amplitude, 2, 61
annealing, 12, 17
attachment, 12, 131
automation, 23

B

beams, 53
behaviors, vii, 3
biosciences, 128
brain, 2, 130

C

chaos, 125, 130, 132
chaotic behavior, 71
China, vii, 131, 133
clustering, 1, 3, 5, 46, 130
clusters, 44, 45, 46, 78, 79, 130
color, iv
communication, 78, 94
community, 80, 81, 83, 87, 89, 90, 91, 131, 132
complement, 12
complexity, 67
configuration, 6, 46, 99
conflict, 130
connectivity, 2
consensus, 129
cost, 3, 129
coupled time-delay systems, 128

D

damages, iv
decomposition, 8, 14, 84
destination, 23, 56
deviation, 107, 108
diffusion, 42, 125
dynamical systems, 125, 126
dynamics, 3, 7, 26, 29, 37, 48, 51, 58, 78, 108, 116, 117, 118, 124, 126

E

efficiency, 127
eigenvalues, 9, 18, 19, 35, 46, 50, 77, 78, 84, 99
engineering, 128
environmental conditions, 23
equilibrium, 7, 67, 68, 70, 71, 72, 78, 79, 80, 82, 84, 87, 90, 132

F

feedback, 67, 69, 79, 82, 83, 90, 91, 128
field theory, 128
financial support, vii, 121
fission, 130
frequencies, 2
frequency distribution, 2

G

Germany, 123
graph, 1, 3, 4, 12, 20, 95, 106, 109, 124
group size, 53, 58, 63

H

heterogeneity, 14, 15, 125
homogeneity, 12
hybrid, 125, 131
hypothesis, 113

I

impacts, 63
independence, 101
inequality, 98
information exchange, 5
information technology, vii
initial state, 1, 38, 53, 58, 80, 108

J

Jordan, 8

L

lattices, 3, 125, 130
Lyapunov function, 1, 20, 26, 36, 50, 57, 87, 95, 97, 103, 110, 111, 113

M

manifolds, 43, 44, 45
matrix, 4, 7, 8, 9, 10, 11, 13, 14, 18, 25, 26, 27, 28, 32, 35, 36, 37, 41, 42, 43, 44, 45, 46, 48, 49, 50, 51, 57, 69, 70, 72, 75, 76, 77, 78, 82, 84, 90, 91, 94, 95, 96, 97, 99, 100, 101, 103, 104, 105, 107, 109, 110, 112, 113, 117
mixing, 124, 125

N

network degree, 17
neural network, 2, 14, 129, 131
neural networks, 14, 129, 131
neuronal systems, 123
neurons, 130
nodes, vii, 1, 3, 4, 5, 6, 7, 8, 11, 12, 14, 16, 17, 18, 25, 28, 32, 34, 38, 41, 42, 45, 46, 47, 53, 54, 55, 56, 58, 61, 63, 67, 68, 69, 71, 72, 73, 76, 78, 79, 80, 81, 83, 87, 89, 90, 91, 93, 96, 98, 99, 108, 109, 117, 131
noise, 14, 61, 62
nonlinear systems, 128
normal distribution, 38

O

optimization, 3, 129
oscillations, 123

P

parallel, 9
parameter, 12, 13, 16, 17, 20, 25, 63, 101, 102
parameter estimation, 25
pathways, 130
permission, iv
physiology, 123
probability, 2, 3, 72, 101, 102
probability distribution, 2
probe, 125
proposition, 44, 82

R

reality, 15
Royal Society, 131

S

scaling, 124
simulation, 2, 53
spin, 123
stabilization, 90, 132
standard deviation, 38
strategy, 20, 23, 24, 32, 33, 34, 36, 37, 56, 61, 67, 68, 71, 72, 74, 76, 93
structure formation, 128
subgroups, 41, 42, 67, 78

survey, 3

T

temperature, 14
topology, vii, 1, 3, 7, 12, 20, 23, 35, 46, 78, 93, 94, 99, 109, 126, 127, 128, 132
transport, 127

U

UK, 123

updating, 27, 28, 29, 75

V

velocity, 14, 94

W

web, 33, 34, 42, 130
WS model, 3